Reflections on

Science and the Media

June Goodfield

American Association for the Advancement of Science

The views presented in this publication are those of the author
and should not be attributed to the Board of Directors, the officers
or other staff members, or the membership of the American
Association for the Advancement of Science.

ISBN 87168-252-4

AAAS Publication No. 81-5

Printed in the United States of America
Library of Congress Catalog Number 81-66420

Copyright © 1981 by the
American Association for the Advancement of Science
1515 Massachusetts Avenue, NW, Washington, DC 20005

To My Friends in Both Professions

Contents

Foreword	vii
Preface	ix
Reflections on Science and the Media	1
Why Communicate at All?	10
The Constraints	15
The Media	16
The Scientists	31
Some Examples	37
Case 1. The Affair of the Painted Mouse	39
Case 2. Pandora's Box	43
Case 3. Rorvik's Baby	51
Case 4. Victims of Myth: The Thalidomide Affair	68
Conclusion	87
References and Notes	103
About the Author	113

Foreword

Whatever differences in method and outlook distinguish science from the media, it is safe to say that both have a passion for independence, just as both exercise growing influence on social change and values. But even as each guards its independence, and fiercely, there is no denying that they depend on each other: science relies on the media to inform the people, while the media relies on scientists for news. This much is clear, but it does not begin to illuminate the strengths, tensions, and flaws that meander through this crucial relationship.

The American Association for the Advancement of Science has a lively concern to see to the public understanding of science and technology. This goes far deeper than an urge towards the selling, or "popularizing," of science. It arises from a perception that the power of science is not neutral to the affairs of states or individuals, but is central to most of the critical choices and outcomes that will be resolved either by informal decision or by default, and that it takes a heap of understanding. It follows, we feel, that science has a heavy responsibility to understand and meet the needs of the media, while, in turn, the media has as much responsibility to understand the methods, discipline, and limitations that accompany scientific discovery, disclosure, and applications.

This volume is not a position paper that sets out to preach an AAAS dogma on science and the media. Rather, it derives from my decision, three years ago, to commission a series of occasional papers by independent scholars on a variety of matters that seem ripe for reflective essays and which have a timely relationship to the business of advancing science. This essay is the first to see the light of day, and it is intended to flush thought, discussion, and debate. Because the topics of scientific and media responsibility are touchy, such an essay is equally likely to surface stress in varying degrees, which is a calculated risk.

In asking June Goodfield to author the essay, rather than either a working scientist or a media practitioner, my choice was neither capricious nor careless. Dr. Goodfield's work was familiar to me, and well-regarded. Her credentials as a scientist are genuine, and her commitment to painstaking and responsible observation of science and its ethical dilemmas is unmistakable in her published work. I was confident that her perceptions of science and the media would be both even-handed and informed, however disputable her opinions might turn out to be. For her, the task has turned out to be more than either of us bargained for, and a less-disciplined author might well have thrown in the sponge. I have no second thoughts as to my choice of essayist.

Publishing this essay is not sufficient in itself. It was worth doing only if it is read widely and discussed by scientists and members of the media, as well as by the growing legion of those whose interests lie in such fields as the history of science, public policy, and journalism. The point of the essay lies not in who did the right or wrong thing in a particular situation where science and the media interacted for better or worse, but in what roads are open on which science and the media can journey together in the future with shared respect, confidence, and a sense of the scale and difficulty of glimpsing the changing face of science for a concerned and preoccupied society.

WILLIAM D. CAREY

Preface

I realized from the very beginning that an invitation to write a critical essay about the mutual and often uneasy relationship between two distinguished professions is an invitation to place one's head upon the block. As I proceeded, an initial feeling of reluctance turned to one of wryness when I sensed that what I was probably doing was uniting the two professions in a new way: in their opinion of the upstart outsider. For I am neither a journalist nor a scientist but hover uncertainly between the two professions; moreover, I grew up in a country with a press tradition that is somewhat different from that in America, though equally honorable. If in the end I did accept the invitation, it was because I believe that in these complex days good communication of science and scientific ideas is more vital than ever before. So I have deliberately raised questions with the aim of provoking debate. Even if these turn out to be too pointed, or my criticisms misplaced, I hope that some problems may be highlighted and some issues clarified.

This essay is most certainly not a definitive analysis of the problems facing those who wish to communicate science. Rather, it is a highly personal document, being my general reflections on the state of the art and of the relationship between the scientific and journalistic professions. I am deeply grateful for the support, encouragement, and advice of Mr. William Carey, Kathryn Wolff, and other staff members of the American Association for the Advancement of Science. I thank them for inviting me to do this monograph and especially for encouraging me to do it in my own style. Too often, the need to ensure "a balanced content" leads to books and articles of such impeccable blandness that the total effect is to smother the issues in a blanket of warm custard. I was given a total and much appreciated freedom. I hope that I have been suitably provocative, but equally that I have exercised reasonable judgment.

Many people have helped me with discussion and ideas during the gestation and the writing of the manuscript; others by applying a much needed critical judgment to the successive drafts. The first draft was written in 1978, and these words remain. But I was able to incorporate a number of revisions in the light of events that overtook me and criticisms that were pervasive and pertinent. In particular, with the lifting of the injunction on *The Sunday Times* over reporting of the thalidomide affair, the full story of this episode and the conduct of the media in relation to it first became clear, both with articles in *The Sunday Times* in 1979 and with the publication of a book written by its reporters. I was therefore able to include a fourth, and very valuable, case study.

One of the best conferences on the topic of this essay has, alas, not been published. It is "Medicine and the Media: Ethical Problems in Biomedical Communication," a symposium presented as part of the 50th anniversary celebration of the University of Rochester Medical Center, 9-10 October 1975. I have had access to the transcribed proceedings and have found them invaluable, to such an extent that had they been published, many of the themes that I touch on would have been already in print. I refer to this conference throughout as the Rochester Conference. I do not know how many copies of the transcript exist, but I hope that one day these proceedings will be published.

Another detailed discussion of some of the problems in communicating science, placed squarely in their historical framework, has just been published: Bernard Dixon's, "Telling the People: Science in the Public Press Since the Second World War," in *Development of Science Publishing in Europe*, edited by A. J. Meadows (Elsevier Scientific Publishers, Amsterdam, 1980). In this essay Dixon too picks up one of my main themes and problems: the question of the role of the journalist as critic.

All colleagues and friends will be able to recognize their own viewpoints and questions: such errors of fact which do exist in this essay are mine alone. But I would like to thank especially Lord Asa Briggs, Mr. Gerald Delaney, Dr. Rae Goodell, Dr. James Hirsch, Mr. Richard Hutton, Mr. John Leonard, Dr. Stuart Marcus, Mr. Howard Simons, Dr. Maria de Sousa, and Dr. Lewis Thomas.

I also want to thank Joy Cull and Betty Main for their help in the typing of my successive drafts, and Carol Zwick for stalwart support in preparing references and putting the whole manuscript into shape.

JUNE GOODFIELD

Rockefeller University
January 1981

Reflections on
Science and the Media

> *There's always wan encouragin thin' about th' sad scientific facts that come out ivry week in th' pa—apers. They're usually not true.*
>   Finlay Peter Dunne "On the Descent of Man"
>   *Mr. Dooley, on Making a Will* (1916)

Intuitive feelings about the tides of history are notoriously untrustworthy, yet one cannot help but feel that those five years in the mid-1970s that culminated in the public debates over recombinant DNA were something of a watershed for the scientific profession. This is so not only with regard to the profession's perception of itself and its relationship to society, but also in the way it is now regarded by individual members of the public and the press. It would be no exaggeration to say that it was a critical time for science, as more than ever before external pressures impinged hard upon a scientific community unused to that kind of attention.

My perception is that many scientists feel hard done by, believing that the real nature of neither their enterprise nor its practitioners has been properly communicated by the media. On the other hand, I find among journalists a strong feeling that, if there was—or is—a problem, its origin lies squarely with the scientific profession, all too many of whose members have been singularly unconcerned, unhelpful, or just plain arrogant when it comes to explaining science to the nonscientist. From my vantage point, with a foot in each camp, I see both judgments laced through with threads of truth and error, both somewhat unfair, and both simplistic.

In this essay I want to analyze some of the reasons for the reciprocal dissatisfactions of both professions and to raise a number of questions that I believe are worth further discussion. These are personal reflections, directed to scientists and journalists as a starting point for debate between the two groups. But this essay is also offered to the general public, for I want to look at science and the media in relation to the society both serve, while focusing on three things: the proper role of the journalist vis à vis science and the public; the obligation of a privileged profession (science) to inform and to cooperate in the process of informing; and the bounden duty of the public to become informed.

As things presently stand, scientists and journalists—and by extension, scientists and the public—are just as likely to be in an adversary relationship as in a cooperative one, and the relationship is quite different from that enjoyed by the scientific profession in earlier years. When, and why, the change?

Without wishing to take too romantic a view of earlier centuries, especially the mid-nineteenth century, in both Britain and the United States there was, I believe, a greater public sympathy toward science then than now. There was certainly a greater scientific literacy, but there were also some significant differences, as Professor A. J. Meadows has pointed out:

> The comparative scientific literacy of the nineteenth century was achieved by a quite different didactic approach from that [which is now current]. First of all, the knowledge imparted was not necessarily the latest discovery related to a science policy, but part of the existing canon of scientific knowledge and belief. Secondly there was far more confidence in the directly relevant character of science to ordinary life. . . . Finally, there appears to be a dearth in the present situation of the characteristic nineteenth century figure of the science populariser drawn from the ranks of science itself.[1]

Coming nearer to the present time, we can even pinpoint one significant shift in the relationship. The media reflect the times, even as they help to shape them, and science journalism in the 1950s, particularly in the United States, had a different character and quality from that which we find in the late 1970s. This point was made explicitly to me by Howard

Simons, a distinguished science reporter of that earlier period and now managing editor of *The Washington Post:*

> The fifties were a "golden age" for science and established an ambience and an image which the scientific profession fervently wished would persist. But in that heyday of American science, when megabucks were so much the main units of currency that scientists could never conceive that the faucets might be turned off, the profession did little to develop a correct working relationship with Congress, or with the media, or with the public. Some scientists, at least, thought that their percentage of the gross national product would always be the same; they tended to promise rather more than they could produce; their congressional testimony was just a little looser than it should have been; and the media played it all up, even at times distorted the promises.[2]

One might add too, that whenever scientists were forced, as they sometimes were, to justify their work, those who could played the "Cold War game" and called upon the existence of Russia and Russian science to justify doing what they wanted to go on doing—namely, to search for new knowledge, untroubled by external considerations.[3]

Simons continued:

> The spokesmen for science in those days were all the sons of Vannevar Bush, of the great scientific push of World War II, of the Manhattan Project and "the bomb." Once the war was over they were not really interested either in solving domestic problems or in bringing science to bear on them.
>
> The science journalists of the fifties and early sixties reflected this tone. There always was a "gee whiz" element in the reporting of science but this really *was* the time of heroes and heroics: the space rockets shot upward; the satellites circled the Earth; the structure of the next sub-atomic particle was inferred and that of the gene deduced; and all this activity was accompanied by cries of admiration from the onlookers.
>
> But despite the heroics, communication between the professions of science and journalism was never outstanding, even though between individual members of each profession it was sometimes very good. The really "pure" scientists rarely talked for that would smack of self-promotion, and scientists (like Dr. Jonas Salk) who had, wittingly or un-

wittingly, got thoroughly entangled with the media were regarded with profound distrust by their colleagues.[4]

Yet as Simons pointed out, within the sixties something else happened, and what began as a ripple of change became a tide of pressure on the scientific community:

> The public's interest in the Space Age and space technology dwindled; there didn't seem to be any imminent danger that the world would blow up; by the time the 150th or so subatomic particle had been found, and the 25th molecular squiggle on the DNA molecule had been discovered, it was apparent that science was going into a period of consolidation. Equally, there were no immediate, nor striking, nor quantitatively significant applications of these discoveries to technology with consequent benefits to the public at large.
>
> This situation was reflected in the media. Science writers get bored just as other people do, and between 1965 and 1967, many of the old established science writers moved away—turning away from science to other aspects of journalism. A good reporter, like a good scientist, goes from peak to peak and knows when he or she is in a trough. Yet as the sixties moved into the seventies, there was developing in society the beginnings of a revival of deep, concerned interest in social problems, ranging from the delivery of health care or alternative energy systems to the problems of the use or misuse of the environment. Society, ready to turn inward, wanted, and expected, that science would do the same. But with few exceptions, science was not ready.[5]

There are several points here. First, the view of history: As someone who has worked in the United States almost continuously since 1959, I found myself in complete agreement with Howard Simons' perceptions. It is true that from 1960 on a number of "public understanding of science" activities were initiated, but most came to fruition only slowly—if at all—and some potentially important activities that needed support could not get it. For example, Eugene Kone, a doyen of science communicators, said to me:

> Sure, the Council for the Advancement of Science Writing started at that time, but it was a creature of the National Association of Science Writers, not of the scientific community, who gave absolutely minimal help. And to this day it struggles to get funds. As for the National Science

Foundation, the pinnacle of their achievement in public understanding of science occurred in the last decade. This year, 1979, marked the first time its funds in this area topped one million dollars per annum. But between 1955 and 1967, this had a very low priority—and this is spoken by someone who knows. I knocked on their door all those years. Moreover, only very few scientific institutions and universities chose to utilize public relations staff in science, and even today the number of universities with such people can be counted on the fingers of two hands.[6]

As science went about its own concerns, legitimate public queries about its technological derivatives were ignored. Commenting on one particularly significant technology, H. Peter Metzger, science writer and syndicated columnist, in an address to the Atomic Industrial Forum, Inc., said:

In the late 1960s your troubles [began]. They began when citizens started asking perfectly proper questions and were treated with disrespect by Holifield's Joint Committee [on Atomic Energy] and a poppa-knows-best attitude by Seaborg's [Atomic Energy Commission] and [a] combination of the two by the nuclear industry.

People like Holifield, Hosmer and Aspinall on the Joint Committee, and their counterparts in the AEC, the Rameys and the Seaborgs, with undisguised contempt, simply denied the public its rightful voice. . . . ordinary people were treated with sneering impatience that made them and others overnight converts to the cause of nuclear power rebels.[7]

One must point out, of course, that after World War II a number of prominent scientists (Edward Condon, Dael Wolfle, James Killian, Solly Zuckerman, C. P. Snow, Lloyd Berkner, to name just a few) did continue both to take an active part in domestic problems and to proselytize—some think to an exaggerated extent—on the potential of science to solve *all* the world's problems. Nevertheless, Howard Simons is right: the professional scientific ethos reflected an almost total commitment to basic research, and the young researcher entering the profession came into an ambience where success meant a total devotion to a narrow intellectual problem.[8]

Which brings us to the third issue, that of popularization. Professor Meadows was perhaps too pessimistic—noted popularizers span the spectrum of time and of science, from

Eddington to Haldane, Oppenheimer, Dubos, Feynman, Commoner and Sagan. Nevertheless, the young scientist who tried his hand at popularization might well place his career and reputation for "soundness" in jeopardy. Margaret Mead is a case in point: her predilection for writing up her research results simultaneously as articles in learned journals and as beautifully written popular books, of which *Coming of Age in Samoa* was the first, certainly did nothing to win acceptance of her ideas by her peers. She did, of course, ignore their disapproval—but this takes a strong will indeed.

Finally, it is also true that there was—and is—little financial incentive for telling the public about science. Even the fifties and sixties, which saw the great curricular reform projects, saw little money or concern invested in efforts to create a scientifically literate public. So we see that both tradition and lack of overriding imperative have helped make the present uneasy relationship between science and the public seem only natural. And this brings us back to the media, the link between the two.

At present, two kinds of things are happening simultaneously in and around science. First, exciting fundamental discoveries continue to be made, as they were through the fifties and sixties. Second, society's interests, in health care, the environment, nuclear hazards, the ethics of recombinant DNA, go beyond basic science and demand sophisticated analysis of "trans-scientific" issues—to use physicist Alvin Weinberg's term.[9] The problems triggered by society's new interests, coming as they do during a period of gross public cynicism about governments and professional elites, pose new challenges to the science reporter as well as to the scientist. These issues bring science and scientists into areas of controversy and into dynamic and fluid situations where there are not, nor ever can be, single experiments which will give definitive answers in the form of irrefutable facts or watertight conclusions. The journalist, too, in reporting these issues, must not only find "the facts" but must develop, unconsciously or consciously, new sensitivities about scientific judgment and reliability and new criteria about the selection of whose views and judgments to cover.

In this regard, I think it is instructive to compare the press in the United Kingdom and in the United States: British journalists accept much more openly the role of advocate than do their American counterparts. But should journalists take on this role? The new trans-scientific concerns make this a crucial question. To anticipate a matter which I will discuss later in detail, this was *the* central question at the Astor Conference in September 1973, a conference triggered by the trans-scientific issues posed by the tragic story of the thalidomide children and attended by numerous concerned European and American scientists, physicians, and journalists.[10]

Unfortunately, in these times of active social change and active social questioning, many scientists have responded as if facing threats rather than challenges. When questioned, they seem to feel they have lost their image and their constituency, and are now unfairly used as the media's latest whipping boys. The situation is exacerbated by the lack of open and accepted lines of communication and by a wariness and lack of trust on both sides. Although a small but articulate group of scientists and science journalists have always operated intelligently together, the band of those "silent men" has remained influential, and recent developments have in many instances only intensified their distrust of the media.

Many reasonable scientists now see themselves impaled upon the horns of a particularly nasty dilemma. Feeling their larger constituency and their image slipping, feeling, too, that for whatever reason, the public does not really understand either the nature of the scientific process or the work of the scientific enterprise, they do appreciate that their right response should be better, and honest, public communication and education about science. Yet I think one scientist at the National Academy of Sciences's Academy Forum on Recombinant DNA[11] said what too many scientists feel: "The public clearly prefers a simple lie to a complicated truth"—a state of affairs which, he clearly felt, the media had done little to alter. Moreover, many scientists believe that too many people in the media *always will* present the public with simplistic stories rather than struggle to explain complicated truths. They believe that the contest between the bare fact honestly presented and the flashy headline, or between a difficult and prolonged controversy and a short but exaggerated story, is an unequal one; given a contest between truth and profit, these scientists say, the truth will always go under. Moreover, as the late Adlai Stevenson remarked, "you will find that the truth is often unpopular, and the contest between agreeable fancy and disagreeable fact is unequal. For, in the vernacular, we Americans are suckers for good news." And when something like David Rorvik's *In His Image: The Cloning of a Man*[12] is promoted as truth, many scientists clearly feel that Stevenson's sentence should well end after the word "suckers."

Nevertheless, we must remember that the involved members of the public have also been changing in the past twenty or thirty years. There are now many more of them, and participation within the democratic Western societies has proceeded several stages further than the mere acquisition and exercise of the right to vote. We live in an increasingly populist society; politicians are having to pay attention to a larger constituency; and the public has become more interested, more organized, more articulate, and more issue-oriented.

I was once asked by a young graduate at a distinguished scientific institution, after a lecture of mine about the recombinant DNA debates, "Who is the public out there, anyway, that you keep talking about?" The answer is that it is not just the intellectual elite, or the pragmatic manufacturer who, in the nineteenth century, had a direct say in the direction of affairs and the expenditure of the GNP; nor is it, as in the immediate post-World War II years, a thin wedge of people who by virtue of their lobbies, their money, their sheer intelligence or their hard work could exert influence at both the state and federal levels. Now the answer to the young scientist is this: the public out there, whose influence is increasingly felt, is made up of people less and less likely to be in sympathy with elitist points of view. The public is the bus driver who takes that young scientist home at the end of his laboratory slog, the lady who takes care of his laboratory animals, the man who sells him his lunchtime sandwich; and most importantly, the public are those whose science teachers are the media.

So the scientist—and also the science reporter—should be trying to reach the semi-skilled, ordinary person who believes in basic virtues and who wants to know, quite simply, what is going on. Actually, as Edward R. Murrow pointed out in 1959, the challenge we face in both science and democracy is the same, for "we live in an age in which intelligence may not be able to simplify truthfully."[13] But it will not do to abdicate at this point, despairing of reaching ordinary people. To reach them represents our continuing and never ending challenge, whether we are concerned with the complexities of politics, as Murrow was, or of science. But what we must beware of, of course, is continuing to promote the simplistic mythology of the struggle between good and evil. Once the myth was that scientists were sages and miracle workers.[14] Now, often through the efforts of well-meaning people who can with great effectiveness present simplistic viewpoints about environment, medicine, or whatever, there is a new mythology in which the scientist is fiend or threat. So like it or not, science can no longer take for granted an unquestioning acceptance nor even a sympathetic hearing. That makes the job of communicating science more necessary, more vital and, because it must be comprehensive, more difficult than at any other time in history.

Why Communicate at All?

Why communicate at all? First, because science is a great human activity and a vital part of our culture, so failing to communicate it would be as absurd as failing to publish novels or show paintings. I find that a legitimate but not a complete answer. Science does affect us all, not only through the intellectual excitement it provokes but through the changing perception of ourselves that it causes and the practical benefits—or disasters—that its application brings. These are the traditional reasons to communicate science, but in our contemporary situation there are other reasons too.

Ideally, the relationship between science and the media should be one of mutual interaction, in which elements of responsibility, accountability, understanding and determining patterns of social change are imperatives for both parties. The problems that these imperatives raise are both continuous and changing, and I believe that scientists and journalists should share not only an attitude toward the truth, but also a commitment to change. On the one hand, the commitment is to changing ideas, which then give us a better knowledge of the Universe and a more realistic perception of our place within it; on the other, it is to change the status quo. I realize that many American journalists disavow any such commitment and vigorously deny that their actions have any such consequences. Nevertheless, the aphorism of a famous American journalist, "Find out what's going on and raise hell," has been a familiar edict, and is actually the first step to change. Anyone who puts serious pen to paper is *always* committed to change in some form. There may still be a few "traditional" journalists around who believe they are simple scribes, doing a dull and neutral job, but there can't be very many of them. (Dullness is, after all, *the* unforgiveable media sin.)

Thus the reasons for communicating science to the public span a whole spectrum, from the values of education and culture to those of politics and government. In addition, even if it were not true that more and more people desire to influence decision-making processes, it would still be *essential* in a democratic society for them to be so involved. The present scale of scientific discoveries—especially in the area of biomedical research, or of energy development—makes oversight by the public essential and examination of the long-term social consequences vital.

Of course, not every one feels that oversight of science should be in public hands, but this is the philosophy on which a democratic society is predicated, and it is especially important when vast public funds are at stake. To achieve this oversight calls for the examination of several things: the immediate impact of scientific discoveries, their general social and human consequences, and the long-term consequences of the resulting technologies. Certain technologies—computer storage of personal data, for example—if not controlled can ensure the authoritarian power of the state and the loss of individual privacy and self-determination. So unless we see and appreciate the effects of technology, we blindfold ourselves and yield to others the determination of our future.

Yet, although science and technology are a vital public interest, public understanding of and involvement with science has remained slight. Why? I conclude that it is not because the scientific and technological issues involved—as opposed to the detailed theories—are difficult to understand. The questions are not ones of scientific *complexity*, nor are most scientific problems all that esoteric. If they were, then the role of the media could be clearly defined as didactic and explanatory. The real difficulty has to do with the uncertainties associated with research and innovation and with their long-term, real-life impacts.

Real-life issues are so much more ambiguous and divergent than the immediate results of tidy laboratory experiments. The divergence comes in the changing context of problems and their solutions, in the varied opinions and projections of the experts and in the multiple possible mechanisms of action and its consequences. Such complexities will always be difficult to handle because there will always be differing opinions about cost-benefit ratios, social values and so forth. It is regrettable, but true, that real-world analyses provide unambiguous answers only for trivial problems.

So one question immediately comes up: whose views should prevail, and how can we separate out the various interests and influences in every situation? Technology assessment, for example, affects us all, but it has been a primary arena for special-interest groups which alone may have the expert knowledge and the cash for lobbying both the public and the government. We need a mechanism for knowing what the various special-interest groups are doing, and why. Moreover, we need a means of laying bare the different value systems that are held by the various advocates of particular courses of action. The Concorde SST, space, oil, nuclear energy, the industrial applications of recombinant DNA—even the agricultural revolutions initiated in the Third World countries—all arose from the special interests of particular small groups. But in the end, the public pays the bill—or reaps the reward—so the public must somehow see that all these diverse interests are balanced by open debate and informed analysis.

The media come in squarely at this point—but just to raise hell? I think not. Raising hell just for the sake of raising hell may not be responsible, nor indeed, always necessary. Columnist Rod MacLeish, commenting on the death of a famous newspaper, *The Chicago Daily News*, asked what we should expect a newspaper to do for us, and he answered that, like a friend, it should gently, or fiercely, point out to us things that we would prefer not to know about ourselves.[15] I agree. Thus it is certainly not the function of the media to be totally passive, to pass on, mindlessly and neutrally, statements by whoever can guarantee you a sale, or whoever is in power, or whoever is most vocal, as American newspapers did in the days of Joseph R. McCarthy, for example, when so much of what the Senator said was published without comment or dissent. There has to be selection and, in passing, we must notice that in deciding what to popularize, the media already impose their own views of science on the public. But there surely also has to be some analysis, and there has not been much of that in scientific matters. But over and above these basics, I think we must ask whether the members of the media should provide the prerequisites for debate, that is, a thorough, independent assessment of scientific discoveries and technologies. And where they manifestly do not do this, should the scientific profession set up institutions and organizations to this end? Because if it is not the media's job to promote or assist the public with scientific analysis and debate, then whose job is it, and how can the existence and values of all interested parties be made explicit? And without this, how can there ever be public understanding and consensus in matters which could literally determine the future of life itself?

There are inevitably great divergencies over this question. At one end of the spectrum are the frankly skeptical, those who say that "the public" cannot exercise any kind of judgment at all, because not only are its members not well informed—they *cannot* be well informed. So matters must be left to experts. So no debate, and therefore media involvement is not necessary. At the other end are those who consider such an attitude both elitist and defeatist. But those at either end must acknowledge another consideration: traditional decision-making—whether in science or politics—has been surrounded by secrecy. Too often, the facts and the rationale for decisions do not get through clearly or completely, so the role of the press in ferreting these out is vital. If in Britain a strong advocacy role *is* played by the press, it is a very necessary counterweight to the prevalent attitudes in Whitehall concerning the low level of the public's intelligence and competence when it comes to the formation of any policy—scientific or otherwise. It also counterbalances a long and deplorable tradition of secrecy in decision-making hallowed through time by consecutive Cabinets and British civil servants. As we shall see, it is probably no accident that Britain has had nothing like the open discussions over recombinant DNA and genetic engineering that have been so prominent on the American scene.

But whether or not it is the function of the media to provoke the debate with all the necessary background information, it certainly *is* the function of the media to expose and examine particular interest groups. In the same way that scholars have laid bare and challenged fossilized ideas, so too, the media should challenge the automatic acceptance of the experts' advice. Dr. John Ziman, Melville Wills Professor of Physics at Bristol University, summed it up in these words:

> It is the wisdom of the pluralistic society to doubt the competence of any authority to choose wisely on behalf of every citizen. It is not so much that they cannot be trusted not to feather their own nests; it is simply that the questions debated . . . are seldom correctly posed. They refer far too much to what is technically possible or technically optimal, rather than what is socially desirable. This is perfectly exemplified by the history of the Concorde project, where the advisory committee seems to have been dominated by engineers and accountants, rather than by potential passengers or by non-passengers residing near large airports. . . .
>
> The proper antidote to the poison of "technocracy" is something like "participatory democracy". . . . The fundamental problems of the unequal distribution of education, experience and responsiblity amongst the voters of a democratic society are ever with us. But I do not believe that we should regard the underlying issues of technological development as inaccessible to the layman who is prepared to take the trouble to inform himself about them.
>
> In this process of education and self-education, the popularising journalist plays a very important part. His job is to bridge the cognitive gap from the side of the layman. . . . The popularisation of science in the mass media is of immense social value. The good technical journalist can acquaint himself with the essential expert opinions and draw attention to the most significant features of a complex argument. If he really knows his stuff, he can distinguish between genuine expertise and bogus authority.[16]

Yet there are, of course, dangers in relying only upon the journalist. Unlike physicians who may feed on the favor of those who buy their skills by playing directly on the desire for reassurance and certainty, journalists fall into the opposite temptation of continually viewing with alarm. If competition for public attention spoils their judgment, what they tell us may be interesting, even highly instructive, but not altogether to be trusted in our search for understanding and decision. Which brings us to the next problem.

The Constraints

News is history shot on the wing.
    Gene Fowler, Skyline (1961)

When distant and unfamiliar and complex things are communicated to great masses of people, the truth suffers a considerable and often radical distortion. The complex is made over into the simple, the hypothetical into the dogmatic and the relative into an absolute.
    Walter Lippmann, The Public Philosophy (1955)

To say that scientists and journalists share a common aim in their respective spheres—the public expression of the truth—is, of course, to state an absolute ideal. Various constraints operating on each profession create conditions which make this highly laudable aim about as abstract as Newton's definition of a body in absolute motion. The constraints arise partly from special interests, partly from external pressures, and partly from professional ethics; these are actually so different that one can easily argue that the fundamental disparity between the environments of journalist and scientist makes the task of each coming to terms with the other well-nigh impossible. When achievements fall short of expectations, or when disenchantments come to be mutual, an examination of the causes generally reveals that it is not venality nor spite that has given rise to the situation, but rather the separate constraints on each party or the separate processes operating within each profession.

We will examine each of these in turn.

The Media

(i) The Printed Word

There have been many assessments of what role a journalist sees himself playing. One of the most useful is that given by the late Nicholas Tomalin, a reporter on *The Sunday Times* (London) who was a part of a great tradition of investigative journalism which produced the Special Projects Groups and the Insight Team.[17] It is worth quoting because it contains both the essence of the job and the dilemma:

> To say that a journalist's job is to record facts is like saying that an architect's job is to lay bricks. True, but missing the point. A journalist's real function—at any rate his required talent—is the creation of interest. A good journalist takes a dull, or a specialist situation, and makes the readers want to know more about it. By doing so he both sells newspapers and educates people. It is a noble, dignified and useful calling.[18]

The crux of the problem so far as our present discussion is concerned lies in Tomalin's phrase "the creation of interest." There are intense market pressures to sell newspapers and then to sell more. The more the media becomes a "mass" media, the greater the temptation not to give facts or create interest but to promote sensation. Now certainly, from time to time, there are new scientific discoveries which positively spill over with interest so that all the reporter has to do is simply record them. The discovery of a new and important vaccine or the successful landing of a man on the moon contain enough excitement by themselves for the news to be self-propelling. But generally it is not like that at all. As Dr. Norton Zinder, a geneticist at Rockefeller University, has said, science is "always newsworthy but seldom news." One must add to that, "always part of a complex web of uncertainties seldom simple enough for a standard news story." Thus the need to create the interest to sell the newspapers may conflict with the more cautious claims of science. This is one constraint.

There are two other problems which make it extremely hard for scientific ideas to get the kind of careful attention and prolonged treatment they properly require. The first has to do with the traditional structure of a newspaper report. British critic and novelist Cyril Connolly spoke of literature as the art of writing that which will be read twice, whereas journalism is the art of writing that which will be grasped *at once*. A news story is structured like this: you can cut off the last paragraph and still have the story; then you can cut off the next-to-last paragraph and still have the story; you can keep on cutting right to the top—to the headline itself and one following line—and you could—in theory!—still have the story. That is basic journalism, but that is not the way scientific journalism can work. In telling a story about science, the reporter must start by building a series of bridges between the reader's understanding and the essential background information. One builds bridge after bridge until finally an understandable conclusion is reached, but if any one of these bridges is cut out, the whole story collapses. So here is the second constraint, built right into the structure of the way the information is conveyed.

The third constraint, also built-in and also true of television, is the need to find the "new" all the time. Scientists must completely understand that the press not only may be uninterested in dealing with anything long term and continuing; they are in fact not set up to do so. This constraint probably explains—just as much as does the desire to expose conspiracies, or deflate the experts, or sell by sensationalism—why such topics as Laetrile and cloning and bending spoons get massive media coverage. They are new, interesting and immediate. They have a sensational element that helps sell newspapers and attract TV audiences. And when collected all together, they make an impact and a running issue—or sore, if you choose to look at it that way. But eventually each topic goes away and is so completely forgotten that when the truth behind the sensational story is finally exposed, it may get no coverage at all.

The media can also create something big one day and kill it the next, no matter what its intrinsic scientific importance. The controversy over Laetrile, the supposed anticancer substance, with the alleged "suppression of crucial data by scientists in the pay of major corporations," is a case in point and one which is instructive since it hit different parts of the United States at different times. Gerald Delaney, the public information officer at Memorial Sloan-Kettering Cancer Center, has told me he would get calls from one part of the country, then another, in turn. One day from Kansas City they would call and say, "This is really big!" Then, a month later perhaps, a bright television reporter from somewhere else would call and say he was going to break the "scandal" wide open. The next week another call would come from San Francisco; a month later, one from Memphis. All the bright reporters in Memphis and Kansas City began at just the same point and moved toward the same conclusion as had those in Chicago and St. Louis the month before. They thought they had found evidence of a real conspiracy involving scientists, bureaucrats and drug companies. This they would, of course, research and expose. But on investigation they found the whole issue infinitely more complicated than expected and the evidence of gross conspiracy elusive to say the least—if it existed at all. So their interest diminished, and then vanished. As a result, such topics get no detailed analysis, partly because they are not evolving at a "newsworthy" rate, and partly because to deal with them properly would require work that no one had yet done—digging deep into situations to find what is the truth and what is not, to find, in fact, the inherent and genuine focus of interest in the story.

The temptation is, of course, to take the shortcuts, and this is often done. There are two kinds of shortcuts: one is simply not to go deep enough or spend time enough to find the "correct" story; the other is to create interest in an irresponsible way, by bending the facts, exaggerating the impact, distorting the consequences, indulging in a spot of free association, even just getting things plain wrong and not caring. There are enough newspapers and magazines that encourage such attitudes to remind one of Humbert Wolfe's little satire:

> You cannot hope
> To bribe or twist,
> Thank God, the British journalist.
> But seeing what
> That man will do
> Unbribed, there's no occasion to.

There have, of course, been some superb science stories in the popular press and some were written in a great hurry. In spite of time and space limitations, *Time* and *Newsweek* have published some outstanding articles: one thinks of Peter Gwynne and Sharon Begley's story on the problem of global temperature and changing global climate;[19] of *Time's* 1969 moon-walk story;[20] or of Peter Stoler's *Time* cover story on hypertension written with Jean Bergerud and Leon Jaroff.[21] Each was essentially complete and accurate in spite of tight deadlines. The hypertension story, for example, was strictly a fast, one-week production. A planned cover story having fallen apart, a *Time* editor requested and got the hypertension story in just seven days—including all of the interviews, research, writing, rewriting, and checking of a cover story on a highly complex and important subject. This was competent professionalism, supported with all the editorial commitment to accuracy that is necessary for an excellent piece.

That brings us to yet another very real, quantitive constraint, that of resources. How many newspapers can afford full-time, qualified science reporters? How many can afford them but choose to put the money and emphasis elsewhere? And finally, how many newspapers and news magazines serve a given locality, and what can each be expected to do? The answer to this last question is not simple: it depends on where you are and who, out in your locality, you think your part of the media ought to be reaching.

Even full-time science reporters on major papers have their problems. I recall that when John Maddox (now once again editor of *Nature*) handed over to Anthony Tucker his job as science writer on *The Guardian*, he had only one word of warning for his successor: "Beware the sub-editor. He is the enemy." The "sub-editor" is, of course, the man on the copy desk who cuts the story down, eliminates the paragraphs as he sees fit and composes the headlines. As it happened, *The Guardian* gave John Maddox ample space into which he poured a number of long and quite superb articles on science and scientific discoveries. Now if Maddox, with all the space at his command, could feel that the sub-editors were a problem, how much more will they be in smaller papers serving a far less cosmopolitan readership? In addition, reporters on smaller papers have problems ranging from those of access to science news—which generally only comes to them via the wire services—to multiple responsibilities and lack of knowledge in any field of science. Also, there probably has been very little in the past or present education of their editors and sub-editors to attune *them* to the difficulties within science and science reporting. So these are also real constraints which we must not ignore.[22]

Yet even if many newspapers carry simplistic or even distorted science stories under headlines or lead paragraphs which may be scientifically incomplete or even outrageous, we—scientists and public alike—should still respond to these stories and encourage the papers to carry more, not less science. The scientific community in particular should try to understand what the pressures are on such newspapers and to offer both long- and short-term help for this particular dilemma. I also conclude that when there are conferences which bring together scientists and science reporters—such as the American Cancer Society's annual Science Writers Seminar—the interests of science reporting might be well served if the gate-keepers—the editors and sub-editors—were invited along with the writers. For over the months and years of their tenure, these gate-keepers wield a tremendous influence by the very process of *their* selections, both in choosing what story will be investigated or reported, and in choosing what in a particular story shall finally see the light of day, in what form, and under what headline.[23]

Yet another factor tending to place constraints on science reporters and newspaper and magazine editors is rather more subtle and, to my mind, much more insidious. It has to do with the pressure of fashion and a desire to shine. We all, scientists and journalists equally, have images of each other's professions which we would like widely accepted. We all also have images of ourselves which we want preserved. We all, as Tom Stoppard emphasized in *Night and Day* (his new play about newspapers and journalists[24]), take ego trips as we go about our professional business. So if in the "golden age" of science the "fashion" in science reporting coincided with an image of an immaculate scientist with pure and undefiled concepts and skills, now the fashion is the reverse—

and just as silly. Certainly one aim of reporting is demystification, but where scandals patently do not exist, now reporters may feel obliged to go out and create them. Daniel Greenberg, editor and publisher of *Science and Government Report*, writing on the aftermath of recombinant DNA in *The New England Journal of Medicine*, said:

> And in the nirvana of contemporary science writing, nothing is accorded greater honor than revelation of suppression in the pure world of science. Small stuff, compared with Watergate, admittedly, but surely better than anything yet turned up on the relatively scandal-free scientific community. In fact, the earliest public meetings on recombinant-DNA research came at the height of the Watergate revelations, and it will be assumed that some of the now-regret-ridden advocates of openness were inspired by the squalor of Watergate to insist on doing business in the open.[25]

To insist on doing business in the open is a very good thing, but if the laurels of scientific reporting will go to the analogs of Bernstein and Woodward, then it is possible that what we're getting is more fashion and less fact, and the public, currently disenchanted both with political promises and with self-serving professional groups, will, no doubt, be more than willing to ride with current fashion.

I think this attitude came out strikingly in the initial kid-glove treatment which both Dr. William Summerlin and Mr. David Rorvik were given by the media, in comparison with the relatively hostile treatment of "establishment" science. (Both Summerlin's "painted mouse" and Rorvik's "cloning" are discussed later in this essay.) Some may consider that this is merely a fair swing of the pendulum and that the earlier hushed and reverential attitude toward science and scientists had to have its reaction sometime, someplace. But thoughtless rejection is very bad both for science and for a society which needs to understand what the scientists really are doing and what are the real implications of their work.

Finally, there is the constraint caused by the sheer pressure both of the volume and variety of scientific events and of news stories having a scientific component. Swings in fashion and mythology notwithstanding, on balance I still see far too much uncritical reliance being placed in "authorities"—those spokesmen for science who may be self-appointed, or who by virtue of being Nobel laureates or heads of institutions may have this distinction thrust upon them. A recent study by Shepherd and Goode[26] showed that the scientific research enterprise portrayed in the media has a significant pecking order. It is presumed that the higher in the hierarchy a person is, the more he or she knows. So to a journalist in search of an eminent quote, such a person is worth significantly more than somebody working lower down, even though the latter may be doing all the research and knows vastly more about the subject.

Shepherd and Goode point out, for example, that the path to becoming a publicly recognized marijuana expert seems to be unrelated either to doing research on the subject or to writing on the subject in those scientific and medical journals that other scientists consider significant enough to cite. It does appear, however, to be related to two other factors: being the head of a health-related agency or institution or being known in a number of areas tangentially related to the subject. In another study, Rae Goodell of MIT found a similar pattern: there is a group of scientific "experts"—the "visible scientists"—whose views are commonly sought on a wide range of scientific matters simply because for a variety of reasons they have been exposed to the press and their names are known.[27] The authors of *The New Scientist* study ask, "Does this matter, this anomaly which is clearly not present when reporters write about athletics or novelists?" Well, yes, it does matter. It makes the scientific process and scientific knowledge seem more authoritarian than it is, and scientific communication seem more divorced from the real practice of the enterprise, than is either real or necessary or good for us.

(ii) *Television*

If there are problems enough for the print media, it is in television, particularly in news and magazine programming, that we see the constraints grossly exaggerated. The problems of the scoop, of market pressures acting on both the reporters and the owners of the stations, of the speed with which the news must be both presented and then superseded

—all are monstrously distorted here. It is as if one is in a theater where the backcloth is constantly being taken down and reassembled while the action continues. The sense of immediacy and urgency provided by television has, of course, advantages as well as disadvantages. On the one hand, it indeed does furnish us with a place where, on an international scale, "we can sing together and grieve together," in the felicitous phrase of *The New York Times* critic John Leonard. But on the other hand, fast moving, predigested news effectively prevents the viewer from going through the process of thought and assimilation that is provoked by the printed word. The fact that television is our single most potent and widespread form of communication, and also the most influential by virtue of the numbers of people reached, only compounds the problem one faces when trying to communicate science, or any complicated topic, *as it really is.*

It is possible to produce good, even spectacularly good, science programs on television, and it is generally acknowledged that the British Broadcasting Corporation has often done so. But these are not news programs, and as will be clear in the comparison which follows, the methods the BBC uses to achieve this end seem well-nigh impossible to do at all widely in America.

The BBC technique is hallowed by a long tradition stemming from the good fortune of having had Lord Reith as its first director general. Avuncular and paternalistic, he would certainly not get away with his attitude these days, even in Britain. But one can be thankful that, from the beginning of his tenure in 1927, he established a principle which basically has not changed. He saw the BBC as having a much wider and more responsible function than that of entertainment alone. There were standards not only of objectivity and truth to be met, but also of seriousness—one might even say highmindedness. And so "Auntie BBC" has shared with those prim, straightlaced spinsters of an earlier century a sense of rectitude, a conscious sense of duty, and a reasoned and accepted morality. The job of the BBC, Reith believed, was to entertain, but also to serve by exercising a deliberate and conscious choice in the quality and the level of the programs. One did not drift to the lowest common denominator or pander to any other pressures, whether political or market. And the terms of the BBC charter rigidly excludes these influences.

So the BBC has produced an enormous range of scientific programs: straightforward natural history films, dramatized biographies of scientists, serials such as Bronowski's *The Ascent of Man*, and the recent spectacular *The Voyage of the Beagle*, showing in situ, as it were, the development of Darwin's ideas. It has also produced controversial programs on, for example, mind control, and the ethics and safety procedures of the drug industry. The BBC record is an enviable one, and many of the U.S. public television's *Nova* science programs are taken from the BBC *Horizon* series.

There are some unique aspects to the BBC system. Responsibility for the choice of programs and their content does not lie in the higher levels of the hierarchy, as it does in the U.S. system. At the BBC, given that the corporation's stated purposes and principles of morality, objectivity and seriousness are adhered to, then the individual program producer or series executive producer wields enormous influence. He—or she—can choose the programs and topics and need only convince the head of the program series that the idea is good and within budgetary limits. Producers will not have to argue with a committee or an executive board, nor will they have to submit the script to the upper hierarchy. A producer will be totally responsible for the content and the impact of the work, and also, it must be said, will receive and have to answer the public criticism when the program is shown. Only when major series, entailing large sums of money, are at issue will people higher up in the structure be involved at all.

Given that the money is in hand, the principal stages in the making of a BBC television program are these:[28] a period of research comes first during which material is collected and its content discussed with those familiar with the subject, in an attempt to assess not only the method of presentation but suitable participants. Next, material is filmed and interviews are taped. The third stage involves the highly creative work of editing, when questions of balance and taste arise. And basically that is all. Some programs—such as the *The Ascent of Man* series or *The Violent Universe*, which occupied almost the whole evening on one BBC channel—call for an exceptional degree of organization, and many people are involved. At the other end of the scale, a production team of three people produces each episode of a current 15-minute "doctor-series" in less than a week. But whether thirteen long programs or one 15-minute slot, the principle on which BBC television works is the same, and on the whole it works well. The process is very similar in ITV—the British independent commercial television group—and I believe it works as well as it does because of the crucial factor of who controls the purse strings.

The BBC operates by license fees collected and transmitted to the corporation through the government in the form of annual, or even longer, grants. The distribution of funds within the BBC is entirely an internal decision, so there can be a guaranteed continuity of people, of viewpoint, and of quality. This is equally true of ITV, where the funds come via the Independent Television Authority as a levy on external sources and by fees from the advertisers who, however, *never* sponsor individual programs.

The situation in America, even within public television, is remarkably different, and the genuine "independence" of even the public stations is only an ideal. In public television those who want to make programs must also raise the wherewithall, so the creative process is typically subordinate to the search for money. Often, the period of research that the development of a program series demands will be broken by a fund-raising drive to secure enough money to write a trial script; then staff will be hired to do that. Once the draft script is written, the staff may be paid off while the producers try to raise money to make a pilot program. If this is successful, the hiring starts again. Then there may be more firing while a sponsor is sought for all, or part, of the series. Much more time may be spent looking for money than producing the program! In passing, it may be noted that the process is not unlike that of raising funds for a long-term research project, and I find it no accident that similar patterns in both science funding and television funding generate similar consequences in the conduct of science and television.

This manner of financing has several disadvantages. The sword of Damocles—the drying up of funds—hangs over U.S. producers, whereas in the BBC there is a guaranteed source of at least *some* long-term funds. Secondly, hiring and firing and rehiring can lead to great discontinuities in the groups whose job it is to produce the programs. This leads to the third weakness: it is asking a lot of a creative person to maintain his or her creativity and enthusiasm through the years it may take to convince someone to pay for the products of that creativity. Finally, for all their avowals to the contrary, those who provide the money often do intervene in strong or subtle ways.

In American commercial television no one escapes the long arm of sponsorship. A sponsor generally buys a film only when it is finished and only if he likes it. If the sponsor has subsidized the project in part in advance but then happens not to like the films produced, the chances of the series being funded again—or even completed—are vastly reduced. The news section of commercial television is still an exception to this, for sponsors do not see the programs before transmission. But now even the independence of the local news may be being eroded. Consultants—"news-doctors" as they are known in the trade—are advising some managements as to which kind of stories generate the greatest interest (generally those of venality or fear) and what kind of personality makes the most successful anchorperson. When formulas are followed, ratings respond.[29] I'm sure most viewers know nothing about these techniques, but they certainly should be concerned. It will be disastrous if the battle for ratings determines even the content of the news, for then the intention of the First Amendment will surely be subverted.

Now, though American public broadcasting defends its independence, nevertheless its producers too feel the impact of the "sponsor," that is, the underwriters who are the sources of program grants. Both the National Science Foundation and the National Endowment for the Humanities have been known to suggest that particular scripts should be "vetted" by outsiders, and both tend not only to feel it appropriate to influence the content, but to encourage self-invited "overseers" at various stages of program development. When the advice comes from the holders of the purse strings, its impact is likely to be substantial, and when joint BBC/PBS programs are undertaken, as they often now are, the pressure will be felt in London as well as Boston or New York.

In all fairness, I must say that intervention is not an inevitable consequence of sponsorship. The Atlantic Richfield Corporation now owns the London newspaper, *The Observer*, and all the screams of anguish that arose when the take-over occurred have been effectively silenced by Atlantic-Richfield's impeccable sense of responsibility and practical demonstration of its deep desire to preserve that newspaper's independence. This corporation is also underwriting a significant portion of a series on the history of the cosmos presently being made for U.S. public television by Carl Sagan (astrophysicist, author, and Cornell University professor) and Adrian Malone (producer of *The Ascent of Man*). Moreover, other U.S. corporations—Exxon and Mobil come to mind—have as their public television representatives former journalists who, rather than compromising the content, are, so PBS producers have told me, often very helpful in the creative process. But the price of creative liberty will be not only eternal vigilance, but also some continuing, guaranteed source of funding, for otherwise talent and ability are diminished or lost.

Financing is one thing. Treatment of the topic is quite another, and good though the BBC films often are, they certainly are not immune to criticism. In television as in newspapers, the need to catch and hold the viewer's attention within the first minute of the program can lead to distortions and exaggerations in the viewer's mind in much the same way as the eye-catching headline leads to the distortions considered earlier. British scientists have often vigorously criticized the BBC's presentations. Robert Reid, author and ex-BBC producer, describes one professor of pathology (whose work had been described on BBC television) who complained that "the public had more to fear from the ability of mass communications media to distort, misrepresent and terrify, than from any of the biological experiments shown in the programme!"[30] Equally unhappy, his co-worker wrote that the film was an awful warning to scientists who attempted "to come into the marketplace of public relations men, mass media, fashion photographers and pop stars."

American scientists may feel even less temptation to enter the world of television than their British colleagues, for in the United States the commercial pressures and constraints are far more intense. On the other hand, it is no comfort to realize that not only is there less temptation, there is also much less opportunity.[31] For, above all, American commercial television is based on the premise of delivering not a program to the audience, but a mass audience to the sponsor, *at once*. If you look closely at the American networks, you realize how difficult it is for them to produce a good science program. Such programs require a lot of lead time and this is rarely possible. There has to be some solid money, and without a guarantee of viewers this may be almost impossible to obtain. For the producer there has to be a buffer from people breathing down his neck while worrying about the sponsor. He has to have, in fact, all the things that, because there has to be an immediate payoff, standard commercial U.S. television seldom can provide. In commercial decisions, taken from the top downward, safety and caution and ratings are the key words, and sponsors underwrite the programs less to impart information than to accrue esteem and attract buyers. One way, apparently, to garner esteem is to avoid controversy. So, as Carl Sagan, a noted popularizer of science on television emphasized to me, those people who are chosen to speak for science on television too often arrive on screen not because they can be guaranteed to do a good job, but rather because they can be guaranteed not to say anything shocking or ill-judged.

Thus, there are lamentably few decent programs about science on American commercial television, and not all that many more on public broadcasting stations, even though there really is a desire on the part of the public to have more good programs about science. Science on commercial television has been, as Sagan said in an article in *The New Republic*, "dreary, inaccurate, ponderous, grossly caricatured, or (as with much Saturday commercial television programming for children) openly hostile to science."[32]

Certainly this is borne out by a recent study of children's science television by Marcel La Follette.[33] As La Follette points out, one doesn't have to be concerned to improve the image of science: it would be enough at this stage to hope for an accurate image of science at all. So far as science for children is concerned, she found that the Saturday morning cartoons and live action shows rely on "inaccurate, exaggerated science characters." Unfortunately, these shows also make up by far the greatest number of viewing minutes for children, and there is little available between shows for children twelve years and younger and the serious adult level of the *Nova* series on public television.

All in all, so far as the commercial networks are concerned, it is a pretty gloomy picture, though there are small areas of brightness which do indicate genuine attempts to remedy the situation. ABC, on its *Good Morning America* show, makes a strong effort with science and medicine; for example, it recently followed couples through pregnancies, going into some detail on the problems of pregnancy and birth, and finishing with a videotape showing the birth of a baby, which the parents and their doctor later viewed together with the TV audience. There have been regular ABC reports on cancer research and therapy, and CBS has been covering these and other aspects of science and medicine. The science sequences on WNBC's News Center 4 (New York City) are another bright spot, as are some of the science-oriented reports on CBS's *60 Minutes* (though some of these have also drawn criticism from scientists).[34] As this essay goes to press I am hearing about a "renaissance" in the popularization of science, but I am not sure how permanent the television renaissance is. CBS may transmit a regular prime-time, half-hour series, *Universe*, with Walter Cronkite as anchorman. The pilot program was sponsored jointly by Exxon and Hartford Insurance and the first producer, Ron Bonn, had a free hand. His goal was, he said, to have a program "that looks like *Star Wars*, but which sounds like *60 Minutes*."[35] Although the sponsors apparently liked the project, by August 1979 it seemed that the whole idea was in jeopardy. At this writing, however, there are plans for twelve *Universe* programs in 1981, with a new producer. It is important to notice that this series, along with *60 Minutes*, originates in the CBS News section and therefore has a degree of independence not enjoyed by directly sponsored programs.

Similar uncertainties operate in public television. WNET (Channel 13, New York) has been planning its own regular science program comparable to *Nova*. Sometimes this idea is on, sometimes off. However, WNET may produce an eight-part series on the brain with NHK (a major Japanese television company) as co-producer, with a budget estimated at $3.5 million. Three quarters of a million dollars has been provided for the project in an unusual joint grant between the National Science Foundation, the National Institutes of Health, and the National Institute of Mental Health. One development on public television that has just begun is a new 13-week series, *3-2-1 Contact*. These programs are directed toward eight- to twelve-year-olds and basically have two aims, according to the research director of Children's Television Workshop, Milton Chen: "To encourage kids to stay tuned in to science," and to stimulate interest in scientific careers among minority youth. Each episode starts with the cheerful announcement, "Science is fun," an assertion which the fast-paced episode tries to sustain. Short on explanations—a survey found that children yawned at these—the shows "encourage positive feelings" rather than teach science. Although the time the networks give to science is still small, these new directions are genuine attempts to present science within the tight constraints imposed by current methods of funding. We can also hope that the current upsurge in network activity in cable TV will lead to filling in some of the gaps in science programming as well as in the arts.

If American commercial and public television does begin to take science seriously, one hopes that there will be significantly improved cooperation between the institutions of science and the television medium, as well as between the individual scientist and the producer. After an initial period of great mistrust and some bad mistakes, the relationships in London on the whole are good, even though recently (December 1980), there has been a monumental row between the medical profession and the BBC production team over a program on transplants and the criteria of death. The BBC Science Consultative Group now meets twice a year and acts as liaison for scientific institutions. British ITV has scientists on its general, educational, and adult educational advisory councils. Thus there is a means for interchange of information and ideas between British television and the world of science and technology which is not tied to particular programs or funding sources. These are liaisons between institutions, however, and it is between working scientist and working producer where mutual understanding and cooperation is most needed. Not only do producers need to understand science in the construction of their programs, but scientists need a real understanding of just how programs are made and what are the limits, the advantages, and the difficulties of the medium.

The Scientists

The constraints acting on scientists are of two kinds: those that are imposed by the basic nature of science and scientific methodology, and those that have been acquired during the evolution of the profession. The first kind of constraints results because scientists are engaged in investigating the unknown, and this is a very chancy business indeed. The error rate is inevitably high, and a sensible scientist naturally prefers not to be caught out on the limb of unsupportable assertions and nonreproducible results. Careful strictures—including first publication of new results only in professional journals and only after peer review—are designed to minimize wasting resources on "wrong" research efforts. All very nice and reasonable, but it results in a second kind of constraint, a strongly negative, frequently unreasonable reaction against "going public" at all. Not only do many scientists have an ambivalent attitude toward the media and public communication in general, but they have an antagonistic, even primitive, attitude toward anyone who undertakes any popularization of science at all.

"Popularization" seems to translate literally in the minds of many scientists into the French *haute vulgarization*. Too many people in the scientific community literally believe that the process demeans both the person who undertakes it and the content of what is communicated. Since, in the eyes of their colleagues, "demeaning" science is precisely what they are suspected of doing, and since professional stakes in the career race are so high, for most scientists it simply isn't worth the risk. I think other factors likely to be involved include the increasingly introverted nature of the profession as it has evolved; a competitive career structure where an initial reputation for scientific "soundness" (i.e., orthodoxy) is the very first thing to cultivate; and a scholarly community separated from the body politic, without an adequate sense of professional and personal responsibility for the impact of its work on society. I also fear that this lack may at times be transmuted into an abdication of moral responsibility, for if there is a higher allegiance, it is not to society but to the profession or the discipline. These attitudes are not unique to the scientific profession but are shared by scholars in other disciplines and by members of other professions also, and are a striking and regrettable part of the total social picture today.

To be fair, however, one must try to understand science's two contradictory canons over public communication. The scientific enterprise depends critically on communication, both formal and informal. Data *must* be communicated, but *not too soon*. Premature publication of scientific information has always been discouraged and for reasons already cited: scientific data are vulnerable, their significance may change with time, or other scientists may fail to reproduce the experimental results. Too much is at stake—people's reputations, the soundness of the scientific method, outside expectations—to let the standards of objectivity and truth slip here.

It is sometimes argued that scientists' distrust of popularization is merely a very proper distaste for those who seek and get more public exposure than their work deserves. But I am not convinced. Basically, good journalists cannot be fooled forever. They will ride along for a while, but they do become sensitized to self-promoters. I think the reason for the distrust is a little more primitive: academics can be singularly uncharitable about each other and extremely unfair, and these are attitudes which are exacerbated in times of great competition.

Lest the reader think I exaggerate this fear of seeming to deviate even slightly from the straight and narrow path of scientific virtue, I offer a quote from the preface of W. R. Klemm's *Discovery Processes in Modern Biology,* a book which is actually addressed to "other scientists, especially younger ones," and to the public as well. The book is fascinating. In it a number of scientists write lengthy personal essays about their work and their lives. But, as the editor describes it,

> The biggest problem in producing the book was in recruiting contributors. Many scientists said that they did not approve of any kind of autobiography, that scientists should be self-effacing. One scientist declined to contribute because he wanted to "reserve the task of biographical writing for my later years." Another scientist declined with the strange comment, "I do not want to participate in this effort to impart to others the fun of trying to take pride in one's workmanship." Very early in my efforts to recruit contributors, I was told by one prominent scientist that this project would ruin my career, that I would make many enemies. My status as a "pure" scientist would be tarnished with the image of being an entrepreneur, an exploiter, etc. "You will live to regret this," he said, almost begging me not to do this dastardly deed. I brooded over that warning for days. And then, I remembered the men who made the first commitments (Drs. Harlow, Rocha e Silva, and Ungar). They had much larger images to protect than did I. They made their commitments without knowing who, if anyone, else had agreed. If they could muster that kind of courage, how could I now turn and run?[36]

Fortunately, other scientists also found the courage to contribute, and as an attempt to integrate science into the culture, and to help us to understand the human processes of science, I find the book invaluable.

I myself am now closely caught up in a similar problem. Following the precept of Nobel laureate Sir Peter Medawar that the only way to understand science and scientific creativity is to peer in at the keyhole, I have been with one scientific group on a daily basis now since 1975 and have thus gained a marvelous and detailed insight into what goes on, in a way that the historian and novelist can only dream about. But publishing what I have learned from my experience in these four years is another matter: "It could ruin careers unless most carefully handled," was the comment that the science editor of *Saturday Review* made to me. So serious might this be that we considered not publishing my book for five or six years, when the scientific community will have completely assessed the work of the group I am observing, and keeping the people and institutions anonymous. This device makes me uneasy, for I like others to verify what I write by reference to the open literature of science. We have delayed publication in part, and it will be over two years from the time I stopped recording the work to the time the book is published.* The delay will allow the work to be put into some scientific perspective, but I am sad that four distinguished scientists who read the manuscript—and who all emphasized that such a book was valuable and needed—advised me not only to delay publication, but to preserve anonymity and to avoid specific names and places. This would, they said, avoid professional risks to the people who so generously gave me this unique opportunity—so I have done this too. A fifth scientist, who is himself involved in the wider dissemination of scientific ideas, admitted the need for the book but seriously questioned the propriety of the direct narrative form and said, "It should be written up as a novel."

So we have scientists uttering pious platitudes about public communication and the need for public understanding of their work and its processes, and of their humanity, while continuing to maintain hypocritical attitudes which may well act to prevent the very thing they are asking for. For when they say, "show that we are human," do they mean only that they want journalists to demonstrate only that scientists are nice, kind, hard-working, and modest? And must the injunction, "Tell the public what we do and how we do it, and how they benefit," always carry the hidden message, "but while you are about it, remember that we may not be very charitable toward the particular person whose work you describe?" Any serious writer feels this situation deeply, and it *is* a constraint. For unless you are totally unscrupulous, you work with one hand tied behind your back, because it simply is not fair to prejudice someone's career—if that is what it amounts to—just to publish your own work. That said, the onus and responsibility is, of course, a dual one, calling for judgment and balance on the part of both writer and scientist. Nevertheless, we writers can rightly ask for charity—if that is what

---

An Imagined World, New York: Harper and Row, March 1981.

it takes—for those whose work has intrigued us, or which we feel has a genuine interest to a wider audience.

If one feels one's sympathy rapidly waning over this first constraint, then one is well-nigh overflowing with sympathy over the other constraints imposed by the very nature of science and the scientific process. These will be familiar to any scientist, but can be briefly illustrated by the following anecdote. Early in this century the city of New York was debating whether or not to allow the horseless carriage to invade its streets. There were many arguments for and against, and one such was advanced by Dr. James Walsh, a distinguished physiologist, as reported in *The New York Times* (7 January 1900):

> Insects have their favorite places for laying eggs. These places are carefully chosen. The butterfly may seem to circle around and alight on any flower; as a matter of fact she goes to one particular flower. These things are never the result of chance. Now, it is a scientific fact that the common housefly lays her eggs in horse manure. In this the eggs flourish prodigiously. With the horse off the streets, the fly *must* follow him. It will not be where the horse is not. Thus a serious channel of infection will be done away with and many lives will be spared. The horseless carriage will greatly reduce the death rate in cities.

Dear Dr. Walsh. How could he have been expected to foresee either the rise in numbers of pet dogs—and the volume of *their* manure—or the increase in speed and number of horseless carriages? Whatever the "scientific facts have shown," prediction has led many a scientist wildly astray. We remember that distinguished physicist Ernest Rutherford asserted that "no practical use would ever come out of splitting the atom"; Sir Richard Woolley, former Astronomer Royal of Britain, described planetary travel as "utter bilge"; more recently, Sir Macfarlane Burnet (who shared the Nobel prize with Sir Peter Medawar for their work in immunology) argued very cogently that, "no practical value has ever come from molecular [biology] research, nor will there be any."

These are cautionary tales indeed, but the germ of our problem is here. So much in and around science is tentative. It is often impossible to state fully, or to anticipate fully, the short-term, let alone the long-term, practical consequences of science. Facts do not always stand up; theories do not always bear the burden put upon them; statistics will be revealing or misleading; ideas have to be modified; conclusions must be qualified; preliminary data are very vulnerable, as are the scientists who publish them; scientists differ vigorously as to what has been established and what has not; and science progresses on a path not unlike that of an inebriated spider. There is a direction of sorts, but many diversions; and perspective, both historical and scientific, is essential.

Thus we see now how mistakes can come from both sides. They come from the scientist who inflates himself with the authority of the "expert" and offers his data and speculations as if from on high, and from the foolish reporter who believes him. Or they come from the careless reporter who has extrapolated a statement or a scientific situation almost out of sight, aided and abetted by a scientist who may indeed be amoral, but who perhaps is only in a hurry. What has to be reconciled is the vulnerability of the preliminary data of science, and its tentativeness, with the journalist's need for speed and hard news. To use the words of pharmacologist Dr. Louis Lasagna of the University of Rochester, "What is needed is some sort of larger social ambience such that the public can handle premature (or tentative) stories with equanimity, without at the same time engendering total cynicism in regard to news stories."[37] The question is, can scientists and journalists work together to do this, or are mutual suspicions so great that this new ambience, too, is likely to remain only an abstract ideal?

Some Examples

I want next to look at four episodes that occurred in the past few years which are of some significance to the foregoing questions. But first there are a number of qualifications to be made. It must by now be clear that throughout this essay I am not taking issue with publications like *Science, Scientific American*, and *The New Scientist* which stand out by virtue of the quality of the writers they employ, the standards of responsible reporting they represent, and the ethics of journalism to which they subscribe. Nor do I deny that there have been, and will continue to be, superb examples of science writing and reporting in such wider circulation magazines as *Time* and *Newsweek* and in some major newspapers. There the doyens of the science-writing profession will continue to receive quite justified acclaim for the quality of the work they do and the standards they have set for the rest of us. What I am concerned with here are the mass media and the audience they reach—that political constituency which I mentioned at the beginning of this essay. Whether we should be trying to reach them is a matter some people may still wish to debate. Maybe we should be concentrating our efforts on that thin wedge of people who are perhaps more effective agents for social and political change and who may already read *Science, Scientific American, The New Scientist*, or perhaps even *Nature*.

But I think this really will not do, and it will not do for several very good reasons. One is that ordinary citizens are increasingly the political constituency to which the Congress is being forced to pay attention, and at present these ordinary citizens are frequently not exposed to the ideas, discoveries, and processes of science in a fair manner. On the contrary, this constituency is exposed, and often quite continuously, to the *mass media's selection* of science news events and how they choose to interpret them. With an increasingly large number of political and quasipolitical groups making their voices and their opinions effectively heard—and at a very basic level—for scientists to step back from this problem would be irresponsible.

Next, it must be recognized that at times the mass media have been the most effective agents for the dissemination of scientific information. For example, the investigative report on *60 Minutes* did more than any other single element to tell us all about the dangers of the chemical insecticide Kepone and the negligence of the company involved toward the community of Hopewell, Virginia. Similarly, in 1965, nearly 1,000 possible contacts of a woman with smallpox were quickly identified, largely through stories in the mass media, and these people were then immunized.[38] An exciting issue and a combination of scientific and journalistic detective work leading to an exposure are ideal media fare. But what we are concerned with here is the day-to-day reporting of science, and the image of science and scientists that the mass media projects.

The four episodes to be examined relate to my main question: What should be the proper role of journalists vis à vis science and scientists? We will see how the media handled these stories; we will look at the scientists' response and involvement; and we will try to judge what lessons or morals might be drawn. The first episode concerns a scientific scandal; the second a difficult scientific-social issue; the third what I and many other people thought was a hoax; and the fourth, an exposure of corporate deficiencies—both scientific and moral —in regard to a commonly used drug. These episodes are the affair of the painted mouse, involving Dr. Summerlin of The Memorial Sloan-Kettering Cancer Center; the issue of safety and recombinant DNA; the book by David Rorvik, *In His Image: The Cloning of a Man*, published by J. B. Lippincott in the spring of 1978; and the thalidomide affair.

Case 1. The Affair of the Painted Mouse

From the point of view of the media, the story of the painted mouse broke first, in April 1974, to Barbara Yunker, medical reporter for *The New York Post*. She was told that a scandal was brewing because of an apparent deception—that a researcher had touched up his experimental mice with a felt-tip pen to make it appear that skin transplants from one mouse to another had "taken." (If true, this would have been a very important medical advance.) Her source she will not, of course, name, but he was not at Memorial Sloan-Kettering Cancer Center and didn't even remember the name of the man involved. After a period of checking around and talking to Dr. Summerlin, Yunker finally interviewed Dr. Lewis Thomas, president of Memorial Sloan-Kettering Institute, and Dr. Robert Good, director of the Cancer Center there. Though after her first articles the Institute was to be criticized by reporters and others for its "silence" and "lack of cooperation," Yunker actually records her experience as "an example of the respect with which I think all reporters who do their work seriously should be treated."[39] According to Yunker, Drs. Thomas and Good, having once mentioned the fact of their distress, and their own personal preference that the story should not be printed, then made a genuinely open and thoroughly fair attempt to be sure that she understood the full details of exactly what had occurred.

Pending an investigation, the Institute gave Summerlin leave for two weeks and quite properly invited the press in only after the investigation was completed. But once the story had broken, the press homed in with the speed and accuracy of mosquitoes scenting a warm and juicy victim. They cannot be blamed for this; it was a thoroughly newsworthy story. It was actually their duty to report it fully, for public finances and public trust were involved. But the question for this essay is, *how* did they report it? Summerlin called his own press conference first, and from then until the appearance of Gail McBride's long, critical study of the whole affair in the *Journal of the American Medical Association*,[40] I believe that most of the press forgot the basic ethics of reporting and the professional standards of their jobs. With the notable exception of McBride, most failed to dig deeply into the situation, and to an extent this was true even of the reporters on *Science*. The press appointed themselves judge and jury. It was, of course, the year of Watergate—a time when, I presume, it was simply assumed that institutions and heads of institutions would be likely to be involved in conspiracy. Seen in this light, Summerlin was able to project himself as a naturally sympathetic figure under enormous pressures, being "done down" by a large organization, and so he became in the public eye. Thus the burden of proof was shifted, and to a degree it remained shifted. It became not a question of demonstrating that the central figure, Summerlin, had—or had not—done wrong; that aspect of the affair was played down. Rather it became necessary for Memorial Sloan-Kettering officials to demonstrate to the press that they were "innocent," even though there was no evidence to suppose that they had done anything wrong, or that they had connived in unethical behavior, or had "done" Summerlin "down," or even knew what was going on. (Whether they *should* have known is a valid question, but that is not the main issue here.)

In the two weeks when Sloan-Kettering's investigating committee was at work, reporters had a field day, building up a case on very superficial grounds. The press came in and talked to the people at the head of the Institute on the one hand and to Summerlin on the other. So they reported, basically, this is what we found out *here*, and this is what we found out *there*, and this is what it all adds up to. Quite a different story emerged to those who took the trouble to dig a little further and talk to the people with whom Summerlin had been working then and in the past. But on the whole, the members of the press failed to dig out the basic issues. They did not check with Gerald Delaney, the public information officer at Memorial Sloan-Kettering, to find out all the issues or the names of others to whom they might find it useful to talk. They just took the official reports, Summerlin's press conference and radio interview, and their own interviews with the main figures in the drama. It must be admitted that from the point of view of a scientific "Watergate," the story was a gem, epitomizing completely Daniel Greenberg's "revelation of suppression in the pure world of science." The atmosphere became infectious, and scientists who had reasons to dislike either Sloan-Kettering or its leadership joined in too. Personal gossip surfaced and was widely used. Gail McBride recalls coming under pressure—though not from her journal—to use totally irrelevant personal data about Robert Good's previous marriages.[41] She did not, but others did. What was most disturbing—and very little indeed was done by the media to dampen it down or refute it—was that most pernicious form of attack, guilt-by-association. One critic, shielding himself under the veil of anonymity, went so far as to say that "The worst Watergate-like thing was that one of the accomplices in the overall fraud [Dr. Good] selected the peer-review committee to investigate, including the most intimate colleague [Dr. Boyse] of his executive vice-president [Dr. Old]. This is like Dean and Ehrlichman investigating their own case."[42] Such comments were unethical, inexcusable and could even be considered libelous. Yet of the large number of reporters who covered the case, I found only *one* who really did the hard work to dig out the facts and background.

Gail McBride, at that time associate editor of the *Journal of the American Medical Association*, along with Barbara Culliton of *Science*, interviewed Summerlin at his home. Later, in conversation with me, McBride recalled being uneasy that day; feeling that there was something more here. "I felt we were being used," she said. As a result she took a courageous decision to hold her story to the point at which it might no longer be topical and only to run it when she had done a lot more work. She would be, in fact, last in print—by some months. But she, and she alone, with careful investigation and bringing sharp judgment to bear on the work, on the circumstances and on the people with whom Summerlin was working, was able to write a superb piece in which blame and understanding, sympathy and sharp judgment, were most fairly handed out to all parties. There was no whitewashing of anybody, and something approaching the truth was out at last.

Yet how many read it—or would even have had a chance to read it? The *Journal of the American Medical Association* cannot be described as a magazine for the masses, and so far had the news gone on, and so fast, that by the time Gail McBride's article came out, four months after the event, the issue was no longer "news," and the article was therefore of no "use" to the wire services or the public press. Considered in these terms, would the truth about this matter ever have been "news" to the press? I doubt it, for there was no incentive except that of truth.[43] But this is precisely the issue, as we will see later in this essay.

Yet the press is fickle in the tone it takes and the judgments it makes. That was demonstrated when later the same year it reported on something similar: the possible tampering with immunological experiments by an undergraduate, Stephen Rosenfeld, who worked with David Dressler, an assistant professor at Harvard University. This episode also concerned letters of recommendation purported to have come from Dressler but reportedly forged, thus raising the possibility that the results of immunological experiments on which the student had been collaborating since June 1973 might also have been tampered with. Attempts by other scientists to reproduce the results were not successful, so a statement of "uncertainty and potential retraction" went out to the journals where the findings were first reported. Although in this case tampering with experiments has not been proved, neither can it be disproved. By contrast with the Summerlin affair, in the press reports there was a factual balance about judgments in the second case which was significantly missing in the first. There was no feeling now that Dressler should have known what Rosenfeld was doing. Dr. Konrad Bloch, Nobel laureate and chairman of the Harvard Department of Biochemistry said, "You cannot protect yourself completely... any of us could have fallen into the same trap."[44] This time, this statement (with which I agree) was not challenged. Maybe the press had shot its bolt, or maybe the personalities involved were neither so colorful, nor so suspect, nor so controversial. But it is the obligation of the journalist to maintain the ethic—doing justice to all situations by digging hard enough and deep enough to bring out the truth. This is the kind of basic professional competence that was exercised in the original Watergate investigation, but in the scientific Watergate which turned out not to be a Watergate the press was not so careful. There was the same "presumption of guilt" that we saw earlier in the way the media handled the problem of Laetrile. A massive conspiratorial cover-up by a medical establishment in the pay of drug companies has remained a recurring theme, and only now does one sense that the burden of proof may be shifting a little.

Case 2. Pandora's Box

The next episode is the story of an issue: recombinant DNA. This genetic engineering technique involves cutting out pieces of DNA from one organism and splicing them into the DNA of another, usually a plasmid, a bacterial virus, to produce a totally new organism. The issue was (and is), can these hybrid molecules be a danger to the public? Much has already been written about this and much more will be written. Historians already have a gold mine of material, painstakingly and comprehensively collected by Charles Weiner and Rae Goodell at the Massachusetts Institute of Technology's Oral History Program. Moreover, three popular articles on the history of the topic and the Asilomar conference were so good that they have won awards.[45,46,47] But the history of the Asilomar conference on recombinant DNA, and its aftermath, is relevant here only so far as the media are concerned. Richard Hutton gave the first good analysis of the role of the press in this event, and his *Bio-Revolution: DNA and the Ethics of Man-Made Life*[48] is very instructive in this regard.

In one very significant way, April 1974 was a decisive month for science. On the 17th of that month eight scientists met to discuss possible ways of handling the hazards that might arise with recombinant DNA and ultimately to decide whether the scientific community should be asked not to carry out such experiments until safety rules had been established. By 18 July of that year, this group had composed a letter to be sent to the professional community (the "Berg" letter[49]) which was both tactful and cautious. It called for a general moratorium on certain types of experiments until attempts had been made to evaluate the hazards. That same day the scientists called a press conference, but they almost certainly did not foresee the press reaction. Most science stories are buried deep inside newspapers, and few ever achieve frontline prominence. But this one did. With the editors and subeditors at work, out poured the headlines: "Bid to Ban Test Tube Super-Germ" (*The Observer*, London); "Genetic Scientists Seek Ban—World Health Peril Feared" (*Philadelphia Bulletin*); "Chilling Theme" (*The Toledo Blade*); "Building Vicious Germs" (*The Washington Star News*); and so it went.[50] Exaggerated as these headlines undoubtedly were, there was one mitigating circumstance: as Richard Hutton rightly points out, the press had been permitted to peep into the inner workings of the scientific community but were not given enough background information as to why or how the scientists had arrived at their unprecedented decision. To say this is to be wise after the event, of course, and indeed there is more than a faint touch of, "If Cleopatra's nose had been a half-an-inch longer, then what might have happened in history?" Scientists being professionally constrained as they believe they must be, it is difficult to see what else they could have done at the time, but it is also difficult to see how the initial reaction would have been different even if full explanations had been given. For if the scientists were concerned enough to raise doubts about the safety of recombinant DNA experimental procedures—sufficiently concerned, that is, to call a moratorium on this work—then the press and the public had no alternative but to take that concern very seriously. With the public knowing as little as it does both of the way scientists think and of the nature of scientific problems, the initial reactions were inescapable: either things were dangerous, in which case all ought to be involved; or they were not, in which case why all the fuss?

At their first press conference the scientists allowed their letter to speak for itself, then had a simple question-and-answer session, and that was that! Hutton suggests that the dichotomy between the press and the scientists was broad but not unbridgeable, requiring only that the scientists educate the press and make available the information which would make their intentions obvious. I am not so sanguine. I agree that the dichotomy was not unbridgeable, but I think the issue had new implications for both professions, for which both were unprepared, and so the reciprocal education was just beginning.

The headlines did two things: on the one hand, they guaranteed that the interest in this story would intensify; and on the other hand, they gave the scientific community a chilling foretaste of the kind and the form of coverage that might continue. Anybody with any experience would have said this was par for the course. Certainly the headlines served to heighten the sensitivities of the Asilomar organizing committee with regard to the press coverage of a conference designed to answer questions about danger. The committee felt an "innate discomfort," as Hutton calls it, about the motives and perceptions of the media—others would call it distrust—and they had already experienced the consequences of the press coverage of the Berg letter. But they also saw that "how the press perceived the conference and its conclusions was almost as important as the conclusions themselves."

The conference was to take place in February 1975, and the planning began in the late summer and fall of 1974. Dr. Maxine Singer, head of the Nucleic Acid Enzymology Section in the Laboratory of Biochemistry at the National Cancer Institute and one of the key members of the Asilomar organizing committee, gave me many of the details. It was a major undertaking, for the conference was to be international in scale and time was very short. In September 1974, at the first planning meeting, the questions were first formally raised about the participation of nonscientists in the meeting. There were two possible categories of such people: lawyers, ethicists and the like, and the press.

So far as a session involving lawyers and ethicists was concerned, initially there were two considerations encapsulated in the scientists' first concerns: given the enormous range of technical discussion to be covered, would there be time in the program and would such a session be anomalous? Once these questions had been settled the next issue was one of proportions: how many such people should be invited? The committee then asked Daniel Singer, a lawyer and husband of Maxine Singer, to be responsible for this session, and to handle the entire matter of nonscientist participation. He was well briefed in the whole problem since Paul Berg, a biochemist at Stanford University and one of the earliest and foremost scientific and moral leaders in the field, had discussed it with the Singers when the issue first surfaced back in 1971. That the participants in that "extra-scientific" session turned out to be all lawyers was an accident. Others—such as Leon Kass, ethicist and physician, and Daniel Callahan, director of the Hastings Center's Institute of Society, Ethics and the Life Sciences—had been invited, but ultimately were unable to come.

However, the story of the press' involvement in the conference is a little more complicated and has been represented in two antithetical ways: the press came in over the fierce protests of the scientists, who tried to ban their participation; *or*, they were welcomed with open arms, and the success of their involvement at the conference not only amply demonstrates this but shows how far the scientific community has moved in this regard. The truth is more complicated—and much more human.

The conference was to take place under the aegis of the National Academy of Sciences, so Howard J. Lewis, director of their Public Information Office, was the natural person to handle the media side of things. In July 1974, after the first press conference, he was approached by Paul Berg and David Baltimore (the latter a Nobel laureate in molecular biology and cosigner of the Berg letter) who said that they wished to limit the press participation to two individuals whom they would select. Howard Lewis protested: this was just not possible, both because of the way the Academy had always operated during its long-standing relationship with the media and because of the "sunshine laws," which require that government agencies do not shield their activities from the public gaze and which the Academy had always honored.

Lewis heard no more about the matter until October when he received a call from Stuart Auerbach of *The Washington Post*. Auerbach had got wind of the proposed conference and, having been told by Berg that all press arrangements would be handled by Daniel Singer, telephoned Lewis to ask why. He also added that if the press were barred, as was being rumored, an injunction would be sought requiring the Academy to give public access to the deliberations of the conference. This was all news to Howard Lewis, who immediately called Singer's office. An unhappy exchange followed over what was essentially a territorial matter—a misconception which several years later seem entirely understandable. Singer had been led to believe that Lewis had been approached over press participation but had said he wouldn't do it. Lewis realized that his statement, "We couldn't do it that way," had been read as "We wouldn't do it at all." After boxing around for a while Lewis went down to Singer's office, and Singer telephoned Berg for a three-way call. Lewis reiterated that there could be neither severe restriction of the number of reporters nor direct personal selection of them because (i) the Academy never operated in that way; (ii) any appearance of excluding the public or the press would be potentially explosive; (iii) there was a precedent in the Pugwash Conferences, where restricting the press had been unsuccessfully tried, with consequences that were not always happy; (iv) the imperatives of the "sunshine laws" were very real ones.

Berg, in his turn, voiced the scientists' legitimate concerns: they worried about the available space at the conference center; they wanted to ensure that proceedings were both adequately and properly covered; their desire to limit the press and to choose the representatives reflected their wish to keep the affair at a manageable level, lest the whole business turn into a media circus. This was why they wanted to choose two individuals.

As the discussion proceeded a compromise was reached and some possible ground rules evolved: numbers would be limited and therefore some selection would be made; staying for the whole conference would be a condition of coming in the first place; the press would be asked to file their stories only at the end of the proceedings; only still photographic coverage would be permitted. To these conditions Howard Lewis insisted on adding one of his own: there would be an open press conference in San Francisco when the meeting was over. So the responsibility for the media moved back to Howard Lewis and all press requests were turned over to him.

Slowly the format built up. Initially he was allocated eight spaces. After checking the room at the conference center, he asked for more. Before he issued the invitations he checked with two officers of the National Association of Science Writers. Did they feel that the preconditions for attendance would be generally acceptable? They both said "Yes," but also said, "Don't quote me." Then having decided not to "advertise" as it were, Lewis made a preliminary selection of those people who had already taken initiatives in the matter by writing or telephoning. He also sought a balance of people and publications—for example, by inviting Michael Rogers of *The Rolling Stone*. This was a happy inspiration indeed, for it resulted in both an award-winning article and a marriage—Michael Rogers to Janet Hopson of *Science News*. Step by step, the numbers built up. Periodically, Lewis asked for more space, and eventually sixteen journalists came.

Yet once the decisions about the press were taken, and it must be emphasized that these were not taken in a mean or grudging spirit, everything worked out magnificently: the scientists took the reporters into their confidence; they gave them every access they asked for; in no way did they attempt to hoodwink, cover-up or have private, off-the-record sessions. The organizing committee genuinely tried to educate the journalists about the nature of the issue, the conference, and the science, and they took all the time that was needed. To begin with, of course, the scientists not only had little idea of what they could expect from the press—they only had fears—but they also had little experience with such potentially explosive issues. Only very few scientists—Dr. Sydney Brenner of the Molecular Biology Research Unit at Cambridge, England, was the most striking example—realized that the basic issue was strongly political as well as technical and was bound to be a public matter. But the only restrictions that the committee finally ended with—in spite of their initial hesitations—were their historic set of ground rules, and these turned out to have been very good requirements. With one small exception they were all honored, though one local reporter did slip in—and out—and filed a story mid-way through the conference.

Several things happened as a consequence of the compromise ground rules. The most vivid publications stayed away, refusing to be constrained by the rules. Next, some of the stories which eventually came out were gems of the journalistic art, and some of them quite rightly received awards. Finally, by the end of the conference the scientists were so comfortable with the press that they hardly remembered their earlier qualms. Indeed, one of them, Dr. Stanley Cohen (a pioneer in the recombinant DNA field at Stanford University and one who had earlier shied away from the press), responded to a charge that public advocates had not been present at the conference by pointing to the press as the participating arm of the public. This remark was, of course, not accurate because journalists were present only as informers of the public, and their job was the accurate dissemination of information. But the important thing here was the demonstration of a comfortable relationship. If the whole process was an education for the journalists and, perhaps even more importantly, for their editors and sub-editors, it was equally so for the scientists. For it was indispensable for the scientists to try to translate their difficult, often incomprehensible language into terms that a journalist could understand, and so, as Hutton noted, one further consequence occurred:

> ... to a public growing increasingly aware of the fragility of its own existence, to a body of professionals adapting poorly to the requirements of a relationship undergoing rapid, fundamental change, and to an issue as emotional as it was real, was added the means *for communication which gave anybody with the desire to learn an opportunity to join in on the debate*, if not on the techniques themselves, at least on the issues that the techniques created. *Informed public scrutiny of science had become a real possibility*. The Salem witch-trials were not going to be re-enacted over the controversy of genetic research; the danger of the majority, driven by ignorance, fear and prejudice, tyrannizing the minority had become remote. With the vehicles of communication available it was now up to both sides to use them. [My italics].[51]

It was, and still is, up to both sides, and where possible, the historic ground rules seem to me to be basically worth retaining for other occasions. Whatever the distortions that, from time to time, were to follow, they were never outrageously gross. Whatever the exaggerations which might come—and they did come—they were never disastrous. For the tone had been set and, I insist, *could always thus be set.* In the end a combination of honesty on the part of one profession and restraint on the part of another had set certain standards which at least established a mutually acceptable base line. Of course, as the controversy developed and political participation became more involved—even tortuous—scientists found that they had to say things not once, but again and again and again, and *still* their views were distorted at times. Exhausting this may be, but that is the nature of the political and democratic process. Once an issue is in the public domain, as recombinant DNA is now, the drama has to unfold itself to the very end, until it is played out or rewritten. This is not to say that, throughout the years since the Berg letter, scientists have continued to be honest either with each other or with the press, or that the press has always been ethical, responsible, and hard-working. Yet, five years later almost everyone is still pleased with the outcome at Asilomar. This includes Howard Lewis, though his pleasure is tinged with a rueful irony as he listens to some of the claims that are now made and recalls the historical development of the episode.

Of course, there are scientists who, long after the event, insist that the concern was overplayed and the conference really unnecessary. There are also some incredibly naive scientists who still feel that the whole matter should have been kept under wraps and the press and public excluded. But most realize that not only was the press coverage necessary but that it was very good indeed. This in great measure was due to a compromise that, although initiated in a situation of tension, was nevertheless reached by civilized people appreciating the legitimate concerns and constraints of others. The model, having been established, is there. The morals, for the scientific community and those who speak for them, are also there. The problem is, of course, that one cannot guarantee that honesty and restraint, or ethics and high moral purpose, will continuously prevail. The path of progress is rarely a straight line. One gains and one regresses, and this brings us to my third example, for it illustrates just this.

Case 3. Rorvik's Baby*

> For the writer of fantastic stories to help the reader play the game properly, he must help in every possible unobtrusive way to *domesticate* the impossible hypothesis. He must trick him into an unwary concession to some plausible assumption and get on with his story while the illusion holds. . . . It occurred to me that instead of the usual interview with the devil, or a magician, an ingenious use of scientific patter might with advantage be substituted. I simply brought the fetish stuff up to date and made it as near theory as possible. . . . As soon as the magic trick has been done the whole business of the fantasy writer is to keep everything human and real. Touches of prosaic detail are imperative and a rigorous adherence to the hypothesis. Any extra fantasy outside the cardinal assumption immediately gives a touch of irresponsible silliness to the invention.
>
> H. G. Wells, Preface to the American edition of his novels (including *The Island of Dr. Moreau*), Knopf, 1895.

The episode of the book *In His Image: The Cloning of a Man* by David Rorvik (published by J.B. Lippincott Co., Philadelphia, in March of 1978) has several dimensions which strike at the heart of the issues relating to science and the media. I do not know of anything quite like this episode in the recent history of science writing and reporting. Whether it was the publication of scientific research, or writing about science which was claimed to be true and not science fiction, up to now the material has been presented so that the truth about the statements could be ascertained, if not directly, then by indirect methods. That is, the scientific facts could be confirmed by the well-tried procedures of experimental verification, or by checking references, or by interviewing the people concerned, or by appeal to established theory. But this book was deliberately constructed so that the full truth could not be ascertained unless the author either produced the principals—which he said he had promised not to do—or acknowledged that the book was a

---

*Much of this episode was developed and written with the help of Dr. Rae Goodell, of MIT, to the extent that I cannot always decide where her contributions end and mine begin. I am most grateful to her.

hoax. The American publisher presented the book as nonfiction, but there was a disclaimer, written into both the author's contract and the book, which in effect tried to absolve the publisher of any responsibility for the truthfulness of the book. (English publishers are forbidden by the Fair Trading Act to print such disclaimers, and in Britain the crucial disclaimer was omitted from the book.)

The general outline of the story is, of course, well known. Hitherto all children known to exist have arrived in this vale of tears by one well-tried, thoroughly reliable, unambiguous method. This book claims that not only is another method possible—which in theory it is—but that in practice it has already been used successfully, and that the child of the method, a bouncing baby boy, is an exact carbon copy of his Dad. The author knows because he helped fix it. Having been contacted by a self-made millionaire called "Max," David Rorvik was persuaded to help arrange a secret laboratory project in some exotic far-off place in which a scientist called "Darwin" extracted the nucleus from one of Max's body cells and used it to create a human embryo. This was planted finally in the womb of a surrogate mother, "Sparrow," and nine months later the genetic replica of Max was born—the first human clone.

Here I'm not going to deal in detail with the questions, does the author's scientific claim stand up, is his professional reputation so solid that we surely *must* believe him, or is the book true or likely to be true, for these matters have been dealt with already in several places: in the article I wrote with Rae Goodell;[52] by the scientific testimony in this country and in Europe; and within the context of an hour-long BBC television documentary film (transmitted May 1979) as well as in an earlier, tough BBC interview (Radio 4, Richard Baker, interviewer), when the author was directly challenged by a number of scientists and by science writer James Wilkinson. Moreover, Rorvik failed to respond to the challenge of *The Sunday Times* in England, which publicly invited him to sue for calling him a fraud and a liar.

When the whole picture is put together (as in the Goodell-Goodfield article; André Hellegers' article in *The Washington Post*;[53] or in Barbara Culliton's review in the *Columbia*

Journalism Review[54]) the circumstance is compelling: Max and the baby do not exist and a human being has not been cloned.[55] But by virtue of the way the book was written, and with the author's refusal to give *any* evidence that could be independently checked, the proof of this contention could not be the irrefutable proof of experimental demonstration—that is, by a complete scientific proof as we commonly understand it. But in these terms there is also no available proof whatever of Rorvik's claim. Moreover, as Bernard Dixon wrote in *The New Scientist*,[56] Rorvik acknowledged this through his failure to reply convincingly to any of his challengers or to provide any independent evidence.

Before launching into the main theme of this discussion, one point must be emphasized: objections to this book do not relate to questions of preserving anonymity; rather, the basic issues are these: Given that the author's factual claims were not verifiable and that the publisher acknowledged that they were not, could there be any justification for publishing a disclaimed, unverifiable, *nonfiction* book that makes such a remarkable "scientific" claim? And secondly, have the interests of science, science reporting, and society been properly served by the book and the subsequent publicity? These are central questions in an episode in which both the author and publisher made their own professional integrity—that is, their adherence to proper standards of ethical behavior—a key issue in the controversy that followed publication.

It is instructive to trace yet again how a news story grows. This one began with an unambiguous advertisement[57] in the trade magazine *Publisher's Weekly*. Carrying no disclaimer, the ad asserted, "A human baby created in a laboratory is now 14 months old." The same ad referred to the forthcoming promotion campaign—a substantial $50,000 advertising budget and a national author tour. The ad, which appeared in mid-February 1978, was spotted by a reporter from *The Seattle Post-Intelligencer*, and via a wire service dispatch the story quickly worked its way to *The New York Post*. There it was picked up, and since such a subject was bound to be of tremendous interest, *The Post* created their news story.[58] *The Post* reporters talked first to Lippincott, but its publicity department said little at this stage. The reporters initially were unable to talk to Rorvik or to Edward Burlingame, senior editor and vice president at Lippincott, who was in England at the time. Nor were copies of the book yet available. So the reporters talked next to Jonathan Beckwith of the Department of Microbiology and Molecular Genetics at Harvard Medical School and a leading member of the radical Science for the People movement; to Liebe F. Cavalieri of the Department of Microbiology and Molecular Genetics at Memorial Sloan-Kettering Cancer Center and a known and very vocal opponent of recombinant DNA research; and to Jeremy Rifkin, author, lecturer and political activist, and co-director of the Washington, D.C.-based People's Business Commission. Rifkin—though not a scientist—is a most vocal critic of science and all things establishment. Each was prepared to talk vividly about the horrors of cloning. Thus *The New York Post* soon had a combination of a subject which was of powerful—even terrifying—interest along with quotable sources who were willing to emphasize the terror: "My God, it's worse than Hitler, a thousand times. It's horrible. Cloning should be stopped—violently if necessary. . . "; and so on.[59]

The news story built up and the flavor of melodrama intensified. Meanwhile, Rorvik, unaware of the impending brouhaha, was on assignment for *Penthouse* and staying at the Americana Hotel in New York. As he told Rae Goodell, he happened to drop by the Lippincott offices and found everyone gathered in an emergency meeting. He was asked to stay in town for a few days and then, since the book's production schedule was being accelerated, to go to Philadelphia to read page proofs. He was also asked to keep out of sight until after the book was printed.

Nevertheless, the next morning he said he found reporters pounding on his hotel door, saying, "We want to talk about the baby." Later, looking through the peephole, he saw that they seemed to have gone away, so he dressed quickly, went down the service elevator and then over to Lippincott where "they seemed to be under siege." Various camera crews were trying to move in, so it was into the service elevator once more and finally to a conference room at Lippincott. Somebody went back to the Americana to collect Rorvik's bags, and he checked into another hotel under an assumed name. Later, while everyone was trying to reach and interview him, he stayed with friends up and down the West Coast. Lippincott arranged his first public appearance, on the *Today Show* of 29 March, just as the book became available. A press conference was scheduled and then canceled because, Rorvik intimated, Lippincott feared for his personal safety! Nonetheless, he agreed to a lecture tour. Another press conference was scheduled, and canceled, in England. He was deluged with offers of radio, television, and press interviews. More than 100,000 copies of the book were printed. The paperback rights were sold for the base-line figure of $250,000, lucrative foreign rights were negotiated, and by 7 May 1978 the book was twelfth on *The New York Times* best-seller list. Though sales of the paperback "never lived up to expectations," according to George Schneider of Pocket Books, nevertheless, at that stage everything seemed rosy.

In America, initially, a few critical voices were raised, scientists apart. Leon Jaroff of *Time* magazine was one such voice,[60], as was Robert Cooke of *The Boston Globe*[61] and Harriet van Horne of *The New York Post*. She returned from Alaska to find her paper running the story on the front page and wrote a scathingly skeptical column on the book and its claim about "an egocentric millionaire with a distaste for normal sex."[62]

But by and large, the media treated Rorvik with kid gloves. Some skeptics, generally scientists, were interviewed and quoted in articles written about the book; so one ultimately could read statements from people such as Beatrice Mintz,[63] who was doing work in closely related areas, and Nobel laureate James Watson, who, several years earlier, had written an article about cloning in the *Atlantic Monthly*. In *People* magazine Watson described the book as a "silly piece of science fiction." Nor should we imply that competent science reporters believed Rorvik's story. But they did not bring investigative resources to bear on the book, the claim, or the author nearly soon enough. On the whole, members of the mass media, especially in radio and television, played along with the author and the publisher. That supposed arch dragon of investigative interviewers, Tom Snyder, host of NBC's *Tomorrow Show*, bypassed the whole issue of the scientific credibility of the book by saying he would leave that to the experts, and thus abrogated the role of journalist as questioner, let alone critic! And to the best of my knowledge he did not provide any experts with equal air time. He allowed Rorvik to tell his own story quietly, of the fantastic claim and the extraordinary circumstances of the cloning, and put him under no pressure.[64] Tom Brokaw, of the *Today Show* (NBC), left Jeremy Rifkin to make the entire running for the story and the possibility of human cloning. Not only did Brokaw fail to probe deeply into the claim and its plausibility, but with unbelievable naïveté let pass Rifkin's image of scientists making unending Xerox copies of people by the mere scraping of a few cells obtained by one brushing up against the other![65] Moreover, he too did not give equal time to scientists or others knowledgeable in the field, and Barbara Culliton's comment on the Brokaw-Rifkin interview, that "it left almost everything to be desired,"[66] could be considered a charitable understatement.

Few in the media—whether general reporters or those dealing with the interface between science and technology—felt it imperative to dissect this story quickly or analyze it deeply. Why not? When I inquired, I heard rationalizations that varied from shrugs of the shoulders to comments along the lines of "Well, publishers do this all the time, so what's new?"; to antiscience sentiments generated by and sustained with suitable quotes from people with vested interests in these attitudes; to shrugs of another kind: "Well, it's so implausible that no one's going to believe it. So why should we have taken it seriously?"

In Britain the claim was taken seriously, and the initiative for a critical examination came first not from the scientific community but from the media—including the popular press, which can be as irresponsible as anyone when they choose, but which were now very skeptical, even to the "yellow journalists." Among the serious journals, *The New Statesman* called the book "a load of old rubbish" and went on to run a small column about how bogus science finally does get its "comeuppance," with digs also at Dr. John Taylor, professor at King's College, London, who had been offering some high-sounding explanation from physics of how Yuri Geller could perhaps have been bending those spoons. In the same article, *The New Statesman* was also happy to draw attention to the full exposure on Italian television of the Geller phenomenon by Geller's ex-manager, Yasha Katz.[67]

The Sunday Times, though, went considerably farther. Phillip Knightley, *The Sunday Times* reporter, examined the issue one week before *In His Image* was published in England. He picked up Dr. J. D. Bromhall's letter to *Newsweek*, which pointed out that Rorvik had written to him for details on his experiments on nuclear transplants in mammalian cells at a time when the cloned baby was supposed to be three months old. So Rorvik should really have been telling Bromhall how to clone successfully. Knightley said that Rorvik was a "liar," and his story ran under the headline, "Why Mr. Rorvik Is a Fraud."[68] A little later, when the book was out and Rorvik had made a public appearance in Britain, *The Sunday Times* ran another column in which it repeated its statements and invited Mr. Rorvik to sue for libel. It then informed its readers that Mr. Rorvik had declined to do so " 'because the book is true,' he said"! *The Sunday Times* then went on to explain that members of the public were entitled to ask for their money back from Hamish Hamilton, Ltd., the British publisher, if they felt that they had been conned.

It is interesting to ask why most journalists in America, especially television journalists, did not put the author or the publisher under any real analytic pressure. Fear of libel suits may be one reason, but the libel law is far fiercer in Britain than in the United States. Inability to cope with complex information is another possibility and the American journalists' reluctance to engage in adversary journalism is yet a third. But the answer one is tempted to give is the cynical one: that certain sections of the media had a vested interest precisely in not doing so, and would rationalize this vested interest in a number of ways: There is the "Greenberg phenomenon"—the fashion of looking for conspiracies in the scientific establishment. There is a challenging mistrust of elites and of professionals and of people in "high places." Max Lerner, in *The New York Post*, would say in his article of 29 March 1978, that he took Rorvik's book seriously, that this cloning business *was* probably true, that there might well be a secret hideaway where experiments had been going on. And he would say this in spite of admitting that "all the qualified scientists have called the feat impossible."

The theory of conspiracy "among the high priests" emerged elsewhere, even in the face of no direct evidence, scientific or documentary, for Rorvik's story. Science writer Judith Randal, in an article in *The Progressive*, raised seriously the possibility that the cries of outrage from the scientific community over Rorvik's book were due to their fear of other public investigations and public control of their work.[69] Having been burned by the recombinant DNA episode, they would be most reluctant to let other issues that might invite regulation come into the public domain. Perhaps, but most of the scientists who spoke out on the Rorvik book—James Watson apart—were in no way involved in the recombinant DNA debate, and it is just possible that their cries of outrage had a simpler explanation: given the state of the art in cloning, the claim must be a lie.

As for probing for the truth on the television networks, it was indeed in no one's interest to probe deeply, for then the entertainment value of the book would surely have been lost. Two facts are very relevant: Madison Avenue mores have come more and more to permeate the world of publishing, and in the end what counts there is the balance sheet rather than the merits of the book. Moreover, the world of TV talk-shows needs fodder and needs it badly, just as the publishers and the authors need big publicity for big sales. Each sustains the other. A book may now be signed up primarily on the basis of the author's potential on the talk-shows, and so between the hammer of ratings and the anvil of profit, truth, integrity and literary merit may be squeezed out.

Both Rorvik and Lippincott, of course, have their own separate justifications for proceeding as they did. Rorvik justified his own refusal to give verifiable information on the grounds that the privacy of the participants in the cloning experiment had to be fully protected, even though he claims that he agreed to cooperate with Max the millionaire only if he were permitted to bring the matter into open discussion. But many articles are written in which an individual's privacy is protected yet the information is still scientifically verifiable. In a statement released by Lippincott, Rorvik further rationalized his stance as follows: "If the drama of human cloning can alert the world to perils far blacker and promises far grander than those embraced by that drama, then I cannot but conclude that humanity will be well served by my having related the events that have been so large a part of my life for the last several years."

Let us now consider the role of Lippincott in this episode. Mr. Burlingame, then vice president of Lippincott and now vice president and publisher of the Trade Department, Harper and Row, gave me a two-hour interview (30 March 1978) as I was gathering information for this book, and he also gave a separate interview to Rae Goodell for our joint investigative article. When the questioning turned to the author's controversial claim and the publisher's unusual practice, we found that Lippincott justified its actions on several grounds. Caught, one suspects, between the Scylla of a potential best-seller and the Charybdis of accepted professional procedures, the publisher was forced to steer a curious course to vindicate its stance. One appeal made was to precedent: many other books have been published which turned out not to be true, or to be true only in part. Another appeal was to the author's integrity and credentials. The conspiracy of a professional priesthood was a third appeal, and this reason was also used to justify Lippincott's failure to go for external review, for they insisted that scientists would be *bound* to reject the book. The publisher's own disclaimer was a fourth appeal. Finally—and most significantly—was the unique appeal that the veracity of the book was not the crucial question—for the issues were so vital and the public so ill-informed! Lippincott hoped a "thoughtful reviewer would be less preoccupied with the issue of veracity, but would say that the book raises fascinating issues that the public should understand."

There were then, it appears, social reasons as well as commercial ones involved in Lippincott's decision to publish. So far as the commercial aspects are concerned, it seemed the book couldn't miss, for if the experts agreed that a human clone had been achieved, this would be truly sensational; but equally if they did not agree, so much controversy would be generated that an enormous amount of attention would be drawn to the book. In either case, obviously, the book would be bound to sell: first, because the very act of publishing in this highly unusual way *created* the interest that the publisher was later to use for justifying publication at all; second, by predicting controversy and avoiding early investigation the publisher set up the conditions for a self-fulfilling prophecy. Having decided that the book could be a bestseller, Lippincott simply set up conditions to ensure it would be.

Both Rae Goodell and I separately asked what Lippincott would do if the book *were* revealed to be a hoax. Provided it was still selling, Burlingame told us, they would call it fiction and keep right on selling it. There have been many other publishing hoaxes, he argued, and such a move has precedent. For example, he offered *The Man Who Wouldn't Talk*, the memoirs of a spy published in 1953 by Random House. The book turned out to be a hoax, but Bennett Cerf went right on selling it. But when I raised this with Donald Klopfer, who founded Random House with Bennett Cerf, Klopfer commented, "There are several significant differences. The author, the publisher, and the Canadian government were *all* hoodwinked by the man involved. We all acted in good faith and at the time there could be no external means of assessment, or review by experts. In [Rorvik's] case there were, but Lippincott chose not to use them. When, three weeks after publication, we heard from the Canadian government, we immediately admitted we'd been made fools of. We did not change the category of the book, nor did we disclaim responsibility—and it didn't sell very well either! Being made fools of—that publishers can stand—that doesn't matter and in that instance no one got hurt."

The list of authors who have made fools of publishers is quite long. The author of *Robinson Crusoe*, Daniel Defoe, gave us the spurious *Journal of the Plague Year*. Perhaps the latest hoax is *Nazi Lady*, purportedly the diary of Elizabeth von Stahlenberg: Ms. Gillian Freeman has recently admitted on "Women's Hour" (BBC) to being the author and that the book was a fake. In between we have had such choice examples as Rowena Farre's *Seal Morning*,[70] a seal version of Elsa the lion (of *Born Free*), only there was no seal. One also thinks of Leonard Lewin's *Report from Iron Mountain on the Possibility and Desirability of Peace*,[71] and, of course, Clifford Irving's *The Autobiography of Howard Hughes*.[72] Some of the publishers involved have not behaved with moral scrupulousness and have exploited the situation. But is an appeal to precedent an adequate excuse?

The consequences of obfuscation by Rorvik and relinquishing responsibility by Lippincott led to a considerable muddying of the waters, and that brings us to the third appeal, the conspiracy of the priests. Given the tenor of the times, it is not surprising that rumors were going around that "Yes, it is all true," and "Yes, there is such a facility, but it is run and financed by the CIA, without knowledge of the government." Rorvik admitted in an interview with me that he knew about these stories. But the situation—and the arguments—are much more subtle than that. Lippincott officials repeatedly emphasized the *social* decisions that underlay their act, for, they said, they asked themselves this question: What kind of book would be worth publishing if one cannot tell if it is true or not? It would have to be an ethical one, of course. In addition, because the general public is not well informed about science, a good science journalist should try to find strikingly new ways to write about such issues as cloning, recombinant DNA, and other forms of genetic engineering. It was, they said, the failure of the scientific community to bring such issues to open discussion that had brought about a situation in which others must perform such a vital service.

It was when I asked, had there indeed not been plenty of public discussion of such issues in the previous two years, that my attention was once more directed to the "priests," and to the desire of the establishment to preserve peace and quiet. There are—it was argued—powerful vested interests that control research, its outcome, the how and when and where. So even if the book were not true, it should be published in order to combat this conspiracy of silence about cloning and its problems. For the job of the publisher, Burlingame emphasized, is to give free flow to ideas, and its primary function is to facilitate their spread, not to filter them. I agree. No responsible publisher would publish anything *known* to be false. But *In His Image* was a "social" book that attempted to bring home to a wider public significant issues which otherwise would not be aired. Burlingame continued to insist: If Rorvik's motives were as he said on the last page of his book—a sincere desire to air the issues at a time when the scientific community wouldn't do this—surely there was justification for publishing such a book even if it turned out to be a hoax?

Next we came to the publisher's disclaimer, those sentences at the front of the book which were used to justify the label of nonfiction while admitting ignorance of the truth. Mr. Burlingame likened these to the cautions offered by the Food and Drug Administration, the warnings that governments cause to be printed on cigarette packets. Certainly there is one similarity: in each case responsibility is shifted. In the case of tobacco, the government shifts at least a moral responsibility in the matter of lung cancer and society's general health while continuing to subsidize tobacco farmers with the taxpayers' money. Lippincott also shifted its responsibility to the public who up to now have known what "nonfiction" meant. There is a significant difference, however, for while the government does indeed support the price of tobacco leaf, it is not actually in the business of selling cigarettes and does not engage in a high-pressure campaign to encourage us to buy the very product it warns us about. Publishers are more than printers: they are purveyors of the printed word; they do vigorously encourage us to buy their product; and they usually accept responsiblity for what goes out under their imprint.

Simon & Schuster, one of four publishers who were approached about Rorvik's book, emphasized this point explicitly. Given their ignorance of the truth of Rorvik's claim, and his persistent refusal to offer them any evidence whatsoever, they felt quite unable to publish in a responsible or convincing manner. They were prepared to go along with the author's scruples provided some evidence was made available to someone. Richard Snyder, president of Simon & Schuster, explained the problem to me:

> It is not unusual for authors to insist on anonymity at times, especially in situations such as actual medical ones

when people really could be hurt by public exposure. But even in these cases the usual practice is to ask the author to deposit with our attorneys evidence of the identity of the people and the claim. This evidence need not be—is not —seen by the publishers, but at least it is available. It gives us a degree of protection—legal protection—and assurance that what we are about to publish is true because evidence exists which can be independently checked if the need ever arises. We offered this option to Mr. Rorvik but he would not even do this. So we felt quite unable to publish. For in order to promote a book with conviction one has to have faith in its contents and where was ours? The problem of the ads also worried us a lot, given Mr. Rorvik's attitude. What could we honestly say? What could we claim that we believed was so? We knew perfectly well that we were turning away a possible million-dollar best-seller. But he wouldn't give us one thing. He wouldn't even go along with accepted procedures. So, we had no alternative but to refuse the book.[73]

Earlier, McGraw-Hill had admitted a similar responsibility for the truth of what goes out under a publisher's imprint. They thought they *did* have evidence that Clifford Irving's book was the authorized biography of Howard Hughes. But when they realized they'd been conned, they decided not to publish and took a hefty loss.

In the end it surely must be asked—and asked generally—if the publishers are not responsible for the integrity of the works they print then who is? Any publisher, whether of books, magazines or newspapers, who wants to be taken seriously, who claims some status for his institution and its wares, who claims, in short, to be considered "professional," does not drift into this twilight world of semi-truth or indulge in sensationalism or scare tactics. The *right* to publish does not imply the *duty* to publish *anything* and then disclaim responsibility, moral or legal, for the consequences. *The very notion of being a professional implies an acceptance of moral responsibility for the consequences of one's work which affect both the other members of the profession and society at large. This is my fundamental point in this whole essay.*

So now we come to the final issue—that of veracity, and whether or not it is irrelevant if *In His Image* is true. In an article in *The Washington Post* (12 March 1978), Judith Randal wrote, "Whether the sensational claim is true or false may actually be irrelevant," because, she suggested, in certain organisms some of the elements for single parent replication already exist.[74] Others may feel that the duty of the reporter was to ascertain the truth *as it applied to humans*. When I raised this matter with Howard Simons, managing editor of the paper, he staunchly asserted, "You're not going to get me to knock Judy Randal," but he nevertheless went on to say that, "Yes, of course, the truth is always relevant, and in their own ways journalists and scientists share a common goal: to seek truth out and state it honestly."

The reader may wonder, if human cloning has not taken place, why are we concerned that Rorvik and Lippincott presented it as if it had? First, I believe that it matters because cloning could have such profound implications. In a situation like this, the worth of the objective fact is magnified, and one should neither explore nor report significant issues lightly. There is also a more general question which relates to the simple matter of truth-telling. I see this whole episode as an example of the selfish, short-sighted attitudes that obfuscation so often attempts to cover. Sissela Bok, author of *Lying: Moral Choice in Public and Private Life*, recently wrote that such "bias skews all judgment but never more than in the search for good reasons to deceive."[75] The reasons to obfuscate that were good for Rorvik, and perhaps for his publisher, were not, I suggest, good for the rest of us. It is interesting that author and publisher were so concerned about their own image, their own professional standing, and that they asked us to believe in *their* good faith. But if for them "truth" can become relative, then subjective, and finally irrelevant, why not for us, too? Why should we believe anything they say or do—and why should they care what we believe about them, anyway? These are in fact not uncommon attitudes today, and the final result will be a totally cynical society in which personal responsibility is rejected and indifference and apathy are rife. Such a society is ripe for propaganda, exploitation, and manipulation, and perhaps we are nearly at this point.[76]

I recall asking both Burlingame and Rorvik if their rationale for the way in which this book was presented wasn't simply an example of the ends justifying the means? The former said, "Yes, I guess you'd be right about that."[77] The latter agreed without hesitation: "Of course," he said, "I am a believer in that practice. I describe myself as a situational consequentialist, which in essence is the same thing. It is the only rational way of seeing life."

In closing this episode, I should point out that, in one sense, we have been here before, although in another sense we clearly have not. Rorvik's "facility" on the small island to the west of Hawaii (South America? Malaysia?) is not unlike that on *The Island of Dr. Moreau*. There the doctor carries out experiments using the latest scientific techniques available at the time H. G. Wells was writing—vivisection rather than cloning. And there Wells extrapolates the science of the times to produce a portrait of Dr. Moreau creating monsters, without a shred of compassion or concern. He has an awful lot in common with that much biopsied millionnaire, Max, who is clearly an equally nasty piece of work, and now, by God, there are two of him!

In the course of my three-hour interview (1 April 1978) with Mr. Rorvik, he mentioned to me, with just the right touch of deprecating pride, that someone had compared him and his book to H. G. Wells. So I asked him, "Do you know *The Island of Dr. Moreau?*" "Yes, yes," he replied, "Absolutely."

What can be concluded generally from the episode of Rorvik's baby? It shows us what a promotional campaign can achieve—but it also highlights the ambience and atmosphere in which science now operates and is likely to operate for some time to come. Mr. Rorvik's book is just one of a whole host of marginally scientific productions, purporting to be factual, but which glide smoothly over the evidence or appeal to emotion or irrationality rather than dealing with solid, proven work and what it promises—or threatens. These marginal productions are successful, not because they tell a good story with responsibility to objective truth and to the public, but because they appeal to—and fuel—the kind of paranoia which lack of information about important matters always tends to produce.

Rorvik's book may not be an end point. Other science writers and other publishers may feel justified in extrapolating in a similar manner. One such article, which I think did just this, was challenged in the *Hastings Center Report*.[78] The article ("George Asmann Died Alone: And Now the First Victim of DNA Research at Fort Detrick is Becoming a Non-Person") appeared in the *Washingtonian* magazine and described in great detail the death of a researcher working in a high-security lab and the subsequent efforts of the National Institutes of Health to deny the death and even the existence of the researcher.[79] In the second half of the article, some seventeen pages in, those who read that far found the following statement: "The preceding is fiction; and being fiction, it takes certain liberties with the truth." When challenged, the *Washingtonian* editor defended the dramatic introductory tale as being "the most effective way to get the reader into a complicated subject." The author, Howard Means, was quoted as saying that, " . . . the debate over recombinant DNA has very little to do with whatever hybrid viruses might crawl out of the lab, . . . what is also at stake in the DNA battle are the limits of knowledge, the unfettered right of scientists to pursue their visions." Unfortunately, as the *Hastings Center Report* said, the subtlety of *that* problem was never mentioned as the author frightened readers into believing that a dangerous virus *had* escaped containment and that the federal government *was* attempting a cover-up.

The *Hastings Center Report* went on to say,

> For a variety of historical, legal, and economic reasons, journalists and the newspapers and magazines for which they write have never been subject to the kinds of professional codes that set limits on the potentially unethical practices of other professionals. In turn, journalists are expected to exercise some discretion and responsibility in their writing, especially on controversial issues. . . . Given the Frankenstein image many people have of scientists and the ready belief that the government is always covering up something, a more judicious approach would have dictated a straightforward and accurate account of DNA research and the genuine concerns it does raise.[80]

There are no limiting and binding professional codes on journalists, although newspapers and broadcasters have voluntary codes (which include the obligation to publish truthful advertisements) in operation throughout their industry. Publishers have not even a voluntary code, though in investigating the Rorvik story, Rae Goodell and I did find a general consensus about standards of propriety and methods of confirmation. Should publishers have an agreed code? Representative Paul G. Rogers (Democrat, Florida), raised this question in the hearing on human cloning which was triggered by Rorvik's actions,[81] and I believe it is a matter highly deserving of discussion.

To return to the Rorvik matter, I think neither the media (including the publisher) nor the scientific community should be really happy with their roles in the affair. The media, especially television, blew it up out of proportion, and too many scientists felt they had no reason to speak out because the issue had very little relevance to the development of science. But it did have relevance both to the image of science and to the public understanding of its applications. Ivory-tower purists who feel that getting on with *real* research is what matters leave a dangerous vacuum into which others will step, with serious consequences for the public understanding of science and, ultimately, for their own work. And scientists must realize that, if they leave the hard work and the battles to others, then they have no grounds for complaint when society fails to understand them and their work. Practical politics and societal acceptance do demand effort and time.

One scientist has faced this. Incensed by the unauthorized use made of his work and the unjustified extrapolations of scientific research in *In His Image*, Dr. Bromhall filed a seven-million-dollar law suit against David Rorvik and Lippincott on the grounds of fraud, deceit, invasion of privacy and libel. The defense pleaded the First Amendment to justify its refusal to answer questions about the existence of the principal characters in the book. The first hearing sustained the case for the first three charges, and at this writing the case is still in progress.

Other scientists would be wise not to stand above the fray. And that brings us to consideration of a tragic affair in which the media performed with the highest standards of concern for the public, while some scientists and their corporate colleagues performed in a manner which I will leave to the reader to characterize.

Case 4. Victims of Myth: The Thalidomide Affair

The thalidomide affair is the story of some 8000 severely handicapped children, in forty-six countries. In Britain it was a story of struggle for moral justice and legal redress, waged by a handful of people, and also a story with overtones of David and Goliath. On one side was a great corporation with massive financial and legal resources resting firmly as a solid pillar of the economic scene. On the other side were some anguished parents and a few outraged journalists. Even though the case dragged on over a time span that was measured by their children's growth from infancy into early adolescence, these parents refused to be bulldozed into accepting totally inadequate help for their deprived children. Equally, the journalists from *The Sunday Times* (London), and their editor refused to withdraw from the struggle until certain basic aims had been achieved: full disclosure of the truth of the matter; acknowledgment of moral and legal responsibility; redress for the families. And they stayed with the issue for an incredible length of time for a newspaper: over ten years. (The Watergate investigations, from the journalists' point of view, lasted only about a year and a half.)

It is fitting that I focus on a book recently published by the *Sunday Times* Insight Team, because the story *in* the book and the story *of* the book are both reflections of a system gone awry. The book is *Suffer the Children*,[82] and it describes the children who were "victims of a dangerous myth": that the tranquilizer offered to a number of patients, including expectant mothers, had no side effects, was nontoxic, and was completely safe in pregnancy. The *Sunday Times* team called the children victims of a myth rather than of an accident since the statements of the safety and remarkable properties of the drug, announced "to the credulous world" by the company, not only were false, but were either known to be false, or would have quickly been seen to be so, had accepted procedures for testing tranquilizers been followed. Society too has been the victim of another myth, minor as compared to the children's problems, but one that surely should be laid to rest. This myth says that the drug companies involved could be absolved from legal and moral responsibility because, at that time, it was not "usual" to do tests on pregnant animals or set up trials with pregnant women. This "cosy view" was summed up by one director of Distillers, the company at the center of the thalidomide affair in Britain: "We went through all the necessary tests at the time. We could not have done more."

The *Sunday Times'* investigators showed several things: not only did Distillers not go through all the necessary tests but took thalidomide on trust from Chemie Grünenthal, the original developer, but over and above the reported side effects that had accumulated, the crucial information—the tests and procedures that showed that drugs can cross the placental barrier and can affect the foetus—was readily available in the scientific literature. In addition, the Insight Team also demonstrated that the "experts—doctors and scientists, called on by the press for their comments on the issues, universally repeated the myth—the scientific literature notwithstanding."

An example of this uniformity was the August 1962 letter from the British Ministry of Health to a parent, Mrs. Pat Lane, which bore a striking resemblance to an article in *The Times* by "A Special Correspondent" (Duncan Burn, the industrial correspondent), which in turn was "amazingly similar" to one published on the same day in *The Guardian*, written by *its* medical correspondent, Dr. Alfred Byrne.[83] All three documents took the same line as the "experts," and absolved the drug companies because, as Dr. Byrne wrote, "pharmacologists the world over did not consider it necessary to try to find out if any new drugs might have such effects."

The similarities in the three documents is understandable: the primary source for all three was Distillers. Only Dr. Byrne made this clear at the time. Duncan Burn was later to agree that indeed he had spoken with the medical men and executives of Distillers and had been given access to "relevant" material in the company's files. The Ministry of Health's information was obtained by its senior medical officer, Dr. E. Conybeare, whose immediate predecessor now worked at Distillers, and who had discussed the matter at length with yet another Distillers' medical advisor, Dr. D. M. Burley. Conybeare assured the Minister of Health that "no animal test existed at that time to test for the properties [the drug] had."

As for the other "experts," those scientists and doctors who had, as is usual, been called on for comment, either they were ignorant of the science and the scientific literature and therefore should *not* have spoken with conviction, or they *did* know the true facts of the situation and were "closing ranks" around the drug companies. With a couple of exceptions they tended to be "visible scientists," those who are well known but not necessarily well informed.[84] I have emphasized this particular aspect at length because it was by exerting pressure primarily at this point that the *Sunday Times* team finally was able to get at the real truth.

The story now documented in exact detail in *Suffer the Children* could not have seen the light of day had it not been for a court ruling. For unlike American journalists, those in Britain were not only battling a large corporation, but were hampered by tough libel laws, an absence of a Freedom of Information Act, and an accepted legal tradition that can put a journalist—indeed, anyone—in contempt of court for commenting publicly on a case once it is before the courts. Only when *The Sunday Times* finally took its case to the European Commission of Human Rights was it able to publish the full story. The reader should turn to that book for a comprehensive, impeccably documented account of most aspects of this affair, whether scientific, medical, legal, moral, journalistic or human.

Here I shall deal primarily with the role of the press, adding two extra dimensions: first an interview with Bruce Page, now editor of *The New Statesman,* who in 1968 took over leadership of the thalidomide investigation as part of the Special Projects unit of *The Sunday Times.*[85] My second source is the discussions at the Astor Conference, which I attended in September 1973. The Conference met in the middle of the journalists' involvement with thalidomide and Bruce Page and other members of the *Sunday Times* staff were present.

Partly as a consequnce of the thalidomide story, many of our discussions focused on the "advocate" role of the press, and these also led to a crystallization of a practical suggestion that I shall mention in my conclusion. First, however, I want to review the thalidomide story in some detail because it highlights the complexity of the problems with which this essay is concerned.

The signals that first indicated a cause for concern flared in Australia. In 1960, Walter Hodgetts, a senior representative of DCBL, Distillers Company (Biochemicals) Ltd. of Britain, persuaded John Newlinds, medical superintendent of The Women's Hospital, Crown Street, Sydney, to try thalidomide as a general sedative. Hodgetts also earlier had left samples with Dr. William McBride, a leading obstetrician with a large practice in Sydney. In the period between 4 May and 8 June 1961, McBride alone delivered three babies, all similarly malformed, with atresia of the bowel. However, since the beginning of that year Newlinds had realized that the rate of congenital malformations at Crown Street was running at three times the national average and near five times that in Australia's second largest women's hospital in Melbourne. On a large-scale map of Sydney he and his wife had been regularly pinpointing the addresses of mothers with malformed children in the hope that some possible associations might generate a clue. Finally only one association stood clear: the areas with clusters of malformed children reflected the presence of the doctors who had tended to refer their obstetrical cases to Crown Street, and though initially the Newlinds felt that the incidence might be just an unlucky statistical run, it was soon clear that it was not.

Meeting McBride after the birth of McBride's third case, Newlinds asked him what he thought was the cause. McBride said he was going to think the matter through and would see him the following week. Over the weekend McBride read up on malformations and bowel atresia, and because of a nagging suspicion, he also read up on drug-induced abnormalities as summed up in a recent CIBA Symposium volume. McBride also had the clinical records of his three mothers to hand, and there found one common feature: none of them had taken any drug during pregnancy except thalidomide. By the time he met Newlinds he was certain that thalidomide would be the cause, but he had no proof. Proof would be one thing, convincing others something else again, and acceptance of responsibility, the final difficult hurdle.

Thalidomide was derived not by a research scientist, but by a pharmacist, Wilhelm Kunz. He was at the time employed not by a drug company with a long tradition in this field, but by Chemie Grünenthal, a new pharmaceutical company born as "an offshoot of soap, toiletries and cosmetics." (Distillers in Britain was similarly a latecomer to the drug industry, having been primarily a spirits and liquor corporation prior to World War II.) Once the new molecule had been made it had sufficient "structural analogies" with the barbituates, to suggest—or so claimed Grünenthal's chief pharmacologist—that it might have "hypnotic activity."

The full story of Grünenthal's work on the drug, its "testing," sales pitch and marketing; and of the warnings Grünenthal received and how it reacted, both immediately and in the subsequent legal cases, has been fully detailed by the Insight Team.[86] When the first reports of side effects—peripheral neuritis—came in, the company did the following:

> It lied when doctors wrote asking if they had heard of this sort of side effect before. It denied all causal connection between thalidomide and peripheral neuritis. It tried to conceal the number of cases that had been reported to the company. It tried to suppress publication of reports about thalidomide-induced peripheral neuritis by using influence and by creating diversion and confusion. It sought to counter critical reports with favorable ones, and to get them, it was prepared to spend money, use influence, and create distortion. It fought to prevent the drug's going on prescription, attacked doctors who advocated this control, and used a private detective to try to discover information that could be used against these doctors.[87]

By 1968 nine Grünenthal executives were on trial in a case brought by the public prosecutor of Aachen, North Rhine-Westphalia, charged with "intent to commit bodily harm and involuntary manslaughter." The case against them said:

> (1) They put on sale a drug that even when taken according to instruction caused an unacceptable degree of bodily

harm. (2) They failed to test it properly. (3) They went out of their way to advertise it as safe when they could give no guarantee that it was. (4) In fact, it caused those who took it to itch, shake, sweat, vomit and suffer peripheral neuritis. (5) When these reactions were reported to them, they first systematically brushed them aside; some of them lied to doctors who questioned them; and when reports became too insistent to ignore, they did all they could to suppress them. (6) The drug caused an epidemic of malformed babies.[88]

After two years and seven months of trial and a financial settlement paid jointly to the victims by the company and the German government, the hearing was suspended. The outcome was ambiguous; even more importantly, the full facts of the tragedy and its root causes were still obscure, and a clear legal precedent had not been established that delineated the responsibility of a drug company to its consumers.

In Sweden too there were many obstacles in the way of help for the children: the Swedish Medical Board, which had authorized the use of the drug, said the tragedy was unavoidable, and the ombudsman cleared it of negligence; the lawyer for the children found that expert witnesses were reluctant to give evidence that could be used against the suppliers, and the drug companies closed ranks nationally and internationally. In Sweden, as in Germany, while a financial settlement *was* agreed to (and the amounts are still not known), once again many questions still remained obscure.

The situation in the United States was different. There was no major thalidomide tragedy, thanks basically to Dr. Frances Kelsey, a medical officer at the Federal Drug Administration, who actually received and processed the application by the company, Richardson-Merrell, for the drug to be sold under the brand name Kevadon. From the beginning Dr. Kelsey *was* concerned with the problems of foetal damage, having worked during 1940 with her husband on the teratogenic effect of quinine. But even more importantly, she had some fundamental questions in mind about the drug's metabolism and stability in the body. In addition she was a sophisticated enough scientist to know the true significance and likely implications of some of the previous animal tests. So the pressures put upon her by the drug company, when added to what she felt was a lack of total honesty in the material they submitted, only intensified her caution. Later, when evidence came to her of the side effect of peripheral neuritis, it only strengthened her resolve to insist on proof of thalidomide's safety.

The legal battles in America were, by comparison with Europe, few, prompt, and effective. Certainly the attitude of the drug company was tough, but so was that of both the lawyers and the courts. There were several cases; in both Canada and America, the issues were debated over the six years between 1969 and 1975; there were jury verdicts; and it is thought that the financial settlements were substantial. Yet the full story of the drug's side effects *still* had not come out. It had not been examined in detail by the press at all for several reasons, one of which, it has been suggested, has to do with the great reluctance of the press to print reports of "unconfirmed" side effects of drugs.[89]

That the story came out in full at all is primarily due to a failure of British justice and to the reactions that this failure provoked. It was in July of 1962 that Mrs. Pat Lane wrote that letter to the Minister of Health asking for help for her own child and about one thousand others. Four weeks later, having had not even an acknowledgement, she called her local newspaper, which carried a story in its next issue, saying that she was planning a test case in court. That same day, Mrs. Lane received the letter from a civil servant at the Ministry of Health which said that before being put on the market the drug had "been properly tested according to the state of knowledge" at the time. It was also this same day that *The Times* and *The Guardian* carried their articles which, as we have seen, also exonerated Distillers—primarily because the source of the article *was* Distillers. By July 1962 members of Parliament had raised the matter in the House of Commons and asked for a public inquiry. The Minister of Health refused then and refused again six months later. The parents now had no other recourse but the law. The first writs against Distillers were issued in 1962, and thus, as *The Sunday Times* shows, "the whole affair was sealed by the laws of contempt of court into a legal cocoon from which it did not emerge until 1977."

The parents, lawyers, children and journalists were bound by the British laws of contempt of court which enjoin complete silence about any case before a court as soon as the writs have been issued, on pain of fines or imprisonment. The laws were designed with the best of intentions: that legal issues should not be prejudged by prior exposure of material, and that the only evidence affecting a case should be that produced in the court, under court rules. But from the first moment the first writ was issued, Distillers was able to take the fullest advantage of these contempt laws and the total clampdown on public debate.

The Sunday Times did not move actively into the fray until 1972, although its connection with the case went back some five years. It was in 1967 that the editor, Harold Evans, had been approached by Dr. Montagu Phillips, a pharmacological advisor to the parents. Phillips was bitter and frustrated with the progress of the case against Distillers and offered *The Sunday Times* 10,000 internal documents that Distillers had been obliged to release to the parents' solicitors so that they could prepare their case. In return for the information, *The Sunday Times* was to finance a book (to be written jointly by Phillips and reporter John Fielding) that would tell the story of "how Distillers bought thalidomide from Grünenthal and marketed it in Britain."

Harold Evans liked the idea. At that time he thought the case would be settled out of court, as in other countries, so there would be no full examination of the tragedy. But he believed that the whole story *should* be told. It was a good story, "one that the public interest required to be told," but more importantly, it had to be told since, unlike other major matters of public concern, there was clearly not going to be any other way of getting the information out. So in January 1968 *The Sunday Times* copied the documents and assigned the Insight Team to the case.

In February 1968 the first of a series of major feature stories about the children and their parents appeared. An out-of-court settlement was still expected, though not yet completed, and *The Sunday Times'* lawyer advised that nothing else could be published until every child's action was finished, or the law of contempt could be breached. So then, Harold

Evans suggested publishing an article based on the information from the German cases alone. The advice of leading counsel David Hirst was sought, and he said, "Absolutely not." At this point *The Sunday Telegraph* did write about the German case, but they omitted the "damning details." Harold Evans, still unconvinced, decided to go ahead, and on 19 May 1968 *The Sunday Times* published "a detailed critical article about Grünenthal" which took up four full pages pages of the paper. The German company immediately sent a telegram, claiming libel and also claiming that to publish the article ahead of their trial was contempt of court. However, there were no counter-moves from anyone in Britain.

By mid-1969, seven years from the first writ, it seemed that publication of the Distillers material could finally be planned, for there were only two test cases and these were about to be settled. But then another group of parents of deformed children decided to sue Distillers. So suddenly there were now three interwoven strands in the problem of publication and contempt of court: the first group of parents that *had* sued; the second group encouraged to sue by the publicity about the first group; and then a third group of peripheral neuritis patients who also decided to sue. So altogether in Britain some two hundred and sixty writs were issued, each new writ requiring a new period of silence until the writ ran its course to settlement.

Public reaction to the anticipated rewards for the children was warm, but at that time no one realized the large gap between actual compensation and actual need. However, a *Sunday Times* editorial, "What Price a Pound of Flesh?," raised the first doubts about the settlements and pointed out that they would be much too low.

By April 1971, a detailed 12,000-word article on Distillers, written by Phillip Knightley and based on their internal documents, was ready for publication, but still nothing could be printed while even one case was outstanding. Four years had now passed since Dr. Phillips had appeared in *The Sunday Times* offices and still Harold Evans was determined to publish. The father of one thalidomide child broke away from the common suit in which Distillers had stipulated that *all* parents had to agree to the terms before it would settle with any one of them. *The Daily Mail* published three stories on the theme that Distillers' attitude was tantamount to "legal blackmail." But when Distillers lawyers complained about this, *The Daily Mail* stopped abruptly.

No one else in the media picked up this particular story for they were all afraid of contempt charges. At this point, *The Sunday Times*, feeling that it was probably time for a new thrust, intensfied its work. The Special Projects unit was now headed by Bruce Page, who took over leadership of the thalidomide investigation and began to dig deep. There were two aspects to their work, so Bruce Page told me:

> We really needed a new approach. Most journalists question authorities and experts, get an average of their views and retail them to the public. This really doesn't get you very far. In fact you can only write something useful if you think through the whole process yourself. You must get yourself into the equation.
>
> So we began to look deeply into two aspects: the biochemical and pharmaceutical, and the legal and compensatory.
>
> With regard to the biochemical and pharmaceutical we concentrated first on one question: were the teratogenic effects of thalidomide forseeable? With almost one voice, all the scientific experts, emphasizing that scientific facts were value-free, answered *that* question in the negative. I was very skeptical. Top guys generally shut up—and never say anything directly if they can help it. Others were clearly closing ranks, and it didn't take long to realize that apparently no one had actually studied the reports and experiments on the reproductive effects of tranquilizers *that were already in existence when thalidomide was brought out.*
>
> So I called Andrew Herxheimer, a leading pharmacologist at the London Hospital,[90] and asked him if he could find me a postgraduate student to study the whole literature of tranquilizers, drugs and experimental work on reproductive effects. We'd pay the person of course, and guessed we'd need them for about four months. Andrew said he'd just the guy and sent over Angus Maconnachie.[91]
>
> Four days after he started, Angus and I met in a pub at the end of the day, and the first thing he said to me was "where does all this come from? Who says there's no history of experimenting on the reproductive effect of drugs and tranquilizers?"

Angus at first thought I was simply just wrong in saying that no tests had been done, but by the time he'd done his work we probably knew the scientific literature better than anyone else.

For a start they found that in 1954—the year that Grünenthal synthesized thalidomide, Wallace Laboratories in the United States published their results for experiments done in the early 1950s in the pharmacology of meprobomate (the tranquilizer Miltown). Tranquilizers were used for long-term therapy and therefore Wallace not only studied the effects of a year's daily administration to rats, but also did a comprehensive reproductive study on mating males and females right through gestation, birth and suckling. The paper in which the work was described was published and available in British libraries at the time that Distillers put thalidomide on the British market.

In 1955 a toxicologist at Imperial Chemical Industries developed a new drug to combat anemia in pregnancy. Imperial also did reproductive studies because, among other reasons, there were numerous studies going back to at least 1917 which showed that birth defects could appear in embryos simply as a result of altering the mother's biochemical balance. A classic paper had appeared in 1948, showing that dyes used against the parasite that causes sleeping sickness also produced birth deformities in rats, at dosages so low as to have no effect on the mother.

Burroughs-Wellcome had run reproductive tests on an antimalarial drug (Daraprim) in 1950; Rhône-Poulenc in France ran them in 1954 on Chlorpromazine, a "major" tranquilizer; Hoffmann-La Roche ran them on Librium, Valium and Mogadon. The papers were published in 1959, and while this was after thalidomide was on the market—but before its teratogenic effects were known—it does weaken the universal claim that "such tests were not usual." Indeed, since 1944 reproductive studies in animals, of drugs intended to be sold for human use, *had been routine procedure* in all Hoffmann-La Roche Laboratories. Other organizations which did routine reproductive studies in the fifties were Pfizer, Lederle, and Smith, Kline & French.[92]

Commenting on the scientists' judgments, Page had several things to say: "Part of the problem is that they were commenting and recollecting about too much information, perhaps. They were not right about the sequencing of the experiments, and this was very important. In any case, if they're going to comment publicly they'd better know the literature." Page and Maconnachie finally knew it far better than the scientists quoted "with authority" on the matter. Thus, as far as the first myth is concerned, the facts are these: it was already known that drugs can cross the placental barrier and affect the foetus; many drugs, including tranquilizers, had been subject to both long-term and reproductive studies; "though animal tests to *prove* thalidomide teratogenic were not in general practice before the disaster, tests *did* exist that would have suggested the dangers to unborn children, because well before thalidomide, it had been established that litter resorption is an indication of teratogenic potential."

Page and his colleagues concluded that had thalidomide been subject to the same routine of tests as Miltown, done by competent experimenters, a decrease in litter size would have shown up. While this would not have proved a teratogenic effect, it would have suggested that thalidomide was not safe for pregnant women and that further long-term studies should be done. So, Bruce Page went on, "To the question, should they have done reproductive studies on thalidomide, the only one respectable, i.e., scientific and moral, answer is 'Yes.' "

So far as the legal aspects were concerned—the second arm of their investigation—this was first handled by another team member. But in the spring of 1971, Bruce Page himself had read two long articles in *Modern Law Review* by an actuary, John Prevett, who had given evidence in a hearing in 1969. Prevett used the thalidomide awards as examples of the need to accept actuarial evidence on the rates of mortality, and he showed that the judge had not only ignored this expert evidence, but had made awards of about half what they should have been. No sooner had Page learned this, and with Prevett's help written a long and devastating memorandum to Harold Evans, than another disturbing fact came to light. The majority of parents with writs outstanding were up for financial deals which would give the children a settlement at about half the 1969 court awards. The combined effect was to galvanize the journalists into renewed action.

Their original feeling, that the main issue was how thalidomide came to be made, was one thing. Given that the expected settlements would be fair, they were willing to wait for the due process of law before publishing. But now, as they recall, there was another issue: "The editor and his colleagues on the thalidomide story realized in dismay that they were about to be silent witnesses to an outrage. It could not be allowed to happen. The complex but normal journalistic objective of analyzing the roots of a tragedy had to be accompanied by a simpler but unique objective—winning more money for the children than the lawyers were going to obtain."[93]

What was needed was a major sustained campaign. Evans, thinking over the issues of moral justice, proposed a two-part effort. The first would be stories by the Special Projects unit about the moral arguments involved over the continual delays in settlement and the amounts offered. This would open the area of personal injury litigation—as Bruce Page described it, the issue of whether "the law could ever find a satisfactory settlement by means of a contest between unequal parties." On 24 September 1972, the paper published a three-page article on the "analysis of the arithmetic of deprivation," accompanied by statistics and photographs. The paper argued that the "law was failing to produce justice for thalidomide children." The headline, "Our Thalidomide Children: A Cause for National Shame," was to become the slogan and theme of the campaign that followed. An editorial entitled "Children on Our Conscience" said, "The plight of the children shamed society, shamed the law and shamed Distillers." *The Sunday Times* gave notice that "in a future article" they would examine how the tragedy occurred.[94]

The article was read by a Labor MP, Jack Ashley, who became deeply concerned. He himself was disabled, having lost his hearing completely after a minor ear operation. He met with the editors and after some discussion promised his full support. This was a vital step for even if there were legal problems for the paper, Ashley could nevertheless transfer the campaign to Parliament. He began a whole round of activity. To back him up the paper reprinted the 24 September article and sent it to every MP, newspaper editor and TV producer in Britain. Others in the media didn't respond, but the law did. The Attorney General's office sent along a warning letter saying that a complaint had been received from the legal advisers to Distillers and the article might constitute contempt.

The Sunday Times, protesting that they had made their case on moral not legal grounds, continued the campaign. The following Sunday they published a further major piece, giving prominent space to letters from parents. More material was published even though there was now a strong prospect of contempt proceedings.

Finally the Attorney General wrote to the editor saying there would be no action on the material already published. *The Sunday Times* was free to argue in print for an improved offer and raise the question of personal injury damage, but if the scope of their attack widened there *would* be an argument. The Attorney General wanted Distillers to see the draft of any new article based on the company's documents. Distillers did see the copy, complained, and the Attorney General said he would go to the High Court for an injunction to stop publication. *The Sunday Times* agreed to hold the new article until the case could be heard by the High Court.

The rest of the story must be quickly summarized. The hearing was held on 7-9 November 1972. Distillers argued that the new article was "a conscious attempt to use the platform provided by the newspapers to pressure the conduct and prejudice the position of Distillers." *The Sunday Times* argued that there were two competing public interests: the administration of justice against the right of the public to be informed, and that the balance should favor open publication in this case. The decision, reached ten days later, found for Distillers, as three judges rejected *The Sunday Times's* argument and banned publication of the article.

But a moral campaign could continue, and now Parliament began to hear the arguments. Jack Ashley lead a motion calling on Distillers to face up to its responsibilities. His theme, as described by *The Sunday Times*, was that this was a great national tragedy, "in which the passage of time instead of healing the suffering [had] heightened it."

The Parliamentary debate and the High Court judgment for suppression of the article finally made thalidomide a national issue. The public interest intensified, fed by both the obstinancy of the company and the law's apparent capacity for suppressing the facts. There was great public outcry, amply covered by the media. One development was that thousands of posters appeared, placed all around the country, on doors, railings and windows, attacking Distillers and calling for a boycott of their products. *The Sunday Times* was afraid these posters would undermine their own public campaign, for this was a transparent ruse against the contempt laws, that is, a paper could reproduce the posters as "news" and thus circumvent the law. The poster campaign did have a wide and very direct appeal, even though no paper reproduced the actual wording.

Now the campaign entered its final—economic—phase, as a few Distillers shareholders began to use their own muscle. They compiled a list of others who owned shares and who would be especially sensitive to public pressure, such as trade unions, churches, and local governments. *The Sunday Times* helped by putting together its own list to publish. Harsh commercial forces gradually built up, and even Ralph Nader appeared on the fringe of the action, advancing the idea of a boycott against Distillers unless compensation equal to that awarded in the United States was offered.

Finally, on 3 January 1973 a previous offer of £5 million was withdrawn and two days later Distillers offered ten annual payments to the children of £2 million/year, via a charitable trust. The figure of £20 million had been a goal for the national campaign—and now Distillers pushed for a rapid settlement. After much "stonewalling," the company at last showed signs of "spontaneity and generosity."[95]

There had actually been two goals for *The Sunday Times* campaign: the future protection of the thalidomide children was first, but the second was to try to gain some lasting benefit for society from the suffering of the families. But even at the time of the financial settlement the story could *still* not fully be told. It seems unbelievable, yet a Royal Commission on civil liability for personal issues, as well as renewed public debates on problems of contempt, were proceeding without benefit of publicly accessible information about how the thalidomide tragedy took place. For the original *Sunday Times* article was still banned, and now another injunction was granted to prohibit publication of *any* material derived from the Distillers documents on the grounds of their being "protected, confidential documents."

At the time of the financial settlement there was some public backlash against *The Sunday Times* campaign. In 1973, the year of the Astor Conference, the Oxford historian A. J. P. Taylor distinguished himself by calling their action a "witch-hunt" that exploited "popular feeling such as the late Dr. Goebbels would have rejoiced to direct—a demagogic clamour, far beyond the bounds of reason." The paper responded that the great public clamor was a "public response that the law itself provoked—as facts were banned—leaving only emotion." And as *Suffer the Children* points out, the Goebbels technique was to tell lies and *The Sunday Times* only wanted to get at the truth.[96] Nevertheless, in 1973 the right of the public to know what *The Sunday Times* had known since 1967 was still unresolved even though Distillers documents were needed for other reasons, too: they could, for example, have been of help in drafting legislation on dangerous drugs and disabled children that had been introduced in the House of Commons.

The injunctions which Distillers had won against the paper were based both on the contempt issue and on grounds of confidentiality of the documents. Now, the paper contested these rulings and the lower court found for *The Sunday Times*. As the whole case for the children was now settled, there could be no danger of prejudice. Still, Distillers appealed to the House of Lords, which reversed the ruling of the lower court. And so matters stood until December 1974 when a committee set up to examine the whole matter of the contempt laws reported on their 18-month deliberations and found that the balance between administration of justice and free speech "had moved too far against freedom of the press." Thalidomide was cited as the leading example.

The findings of the committee certainly opened the door but real reform has yet to come. *The Sunday Times* finally took its case to the European Commission of Human Rights. The final hearing in this whole saga was in July 1977 before the European Commission, which found in favor of *The Sunday Times* with the argument that "there was a higher public interest in learning how a disaster had been inflicted on innocent people."

So on 21 July 1977 Phillip Knightley's original, 12,000-word article, which had been written and ready since April 1971, was finally published—four years and nine months since the original ban, and fifteen years after the first thalidomide child had been born in Britain. *The Sunday Times* says this was the first time "any light at all had been shed on what went on in Distillers during the manufacture and marketing of thalidomide; but more important than the incidental dent in the British law of confidentiality, was the Commission's reasoning in declaring that the British law of contempt in the thalidomide story violated rights to free speech.... *The Commission (in arriving at its decision) said that it had taken into account an additional element in the case—the role of the press in a democratic society and the duties and responsibilities of individual journalists.*"[97]

A famous victory? Well, yes, it may be. In December 1979, Harold Evans, editor of *The Sunday Times*, was presented with the gold medal of The Institute of Journalists. Evans was only the sixth journalist and the second Britisher to be given the gold medal, and the presentation cited the verdict of the European Commission of Human Rights as a famous victory for the freedom of the press. And recently Lord Scarman, Lord of Appeal (the British equivalent of a U.S. Supreme Court justice), in comments made in a foreword to a new booklet on the thalidomide case, said that the case for reform of the English law was a strong one, and the case for incorporating the European Convention into English law was now unanswerable.[98] But the final victory is yet to be achieved, for, unbelievably, in 1981, the English laws, so far as contempt is concerned, remain exactly as they were, nor has any progress been made on a Freedom of Information Act.

One single and most intriguing consequence of the thalidomide affair stands out in my mind. There were many magnificent elements, of course: penetrating investigations; battles for principles such as freedom of speech and the right of the public to know; tenacity over an unprecedented period of time; the willingness of a newspaper to lay out its own money in a long-drawn-out legal case. But over and above these was the willingness to do more than act as a purveyor of information. For in the end the journalists on *The Sunday Times* did much more than lay out bare facts, point out discrepancies, or recommend action. They took a definite moral stance. And their right to do so, and the necessity to do so, are, I believe, crucial factors to be debated in the ongoing relationship of science—and technology—and the media.

"We knew very well what we were doing," said Bruce Page when I raised the matter with him. "With our final headline we were indeed making a moral judgment."

"Should journalists do this?" I asked him. "And doesn't this extend their traditional role?"

"Yes, I think they should, and yes, it does," he replied, and then amplified the point further.

Today, he said, in either political or science reporting, there is no longer a hard and fast line between the facts and their implications, no clear-cut point where the one becomes the other. The theory that a journalist can be totally objective is as much of a myth as the same theory applied to a scientist—"the bucket-theory of absorption," as Page called it. But actually life is a matter of perception and this affects what one sees and how one acts. The thalidomide affair demonstrated this well. "We saw," Page went on, "the actual way in which the British law screwed up the issue of compensation for the children, and we felt we had to make static about that. We also saw that the facts of the case could not be neatly separated from the moralities of the case, and we were prepared not only to say this but to go along with the implications."

Perhaps this additional—and at that time unexpected—dimension of the journalist's role partly explains some of the outrage—or was it fear—that *The Sunday Times*'s role generated in some people. There was, as we have seen, a backlash which varied from feelings that one shouldn't say nasty things about a major drug company to accusations of indulging in a witch-hunt. But there are certain distinctions that must be made here between the nature of facts, the implications of facts, and any consequent actions that are called for.

"If we showed," Bruce Page said, "that in this unknown country there was a real tiger in the bushes and not just innocent children playing, it was nevertheless not our job to provide society with a gun. You go to gun-smiths for that."

Now when, at the Astor Conference, we discussed hard and long the traditional versus the expanded role of journalists, our conclusions were reasonably unanimous, though not arrived at easily. Finally, scientists, doctors, and journalists from both press and radio and television felt that there was no way that this additional role could be avoided, nor should it be avoided. But it could not be taken on by your average daily reporter, working to a deadline. The need for such a role would surface not only in the reporting of science, but in many areas—politics, business, law, and others. And where issues arise that called for a role of this kind, it is the bounden duty of the press to see that the journalists responsible are well versed in the issues, educatable about the facts, and willing and able to spend the necessary time to do the job properly. Without such safeguards the consequences of an extended role for journalists could be as dangerous to society as their silence.

Conclusion

> "What the mass media offer is not popular art, but entertainment which is intended to be consumed like food, forgotten, and replaced by a new dish."
> W. H. Auden. "The Poet and the City." *The Dyer's Hand* (1962)
>
> Guthrie: "People do awful things to each other. But it's worse in places where everybody is kept in the dark. It really is. Information is light. Information, in itself, about anything, is light."
> Tom Stoppard, *Night and Day* (1978)

There are two facts that must be emphasized again. First, the newspapers and television are, and are likely to remain, imperfect instruments for the communication of science and scientific ideas. In fact, they are imperfect instruments for the communication of anything—as Tom Stoppard beautifully demonstrates in his play, *Night and Day*. Jules de Goncourt once wrote that that sheet of paper, the newspaper, bore the same relationship to a book as a whore did to a decent woman. That was too harsh by far, but certainly there are many scientific stories and issues that, given the nature of the constraints upon the media, just must be handled speedily and not in a continuously developing relationship as in a book. It is not as if the ethics of good journalism are not there, but in the hurry of days they may be ignored, conveniently bypassed, or just forgotten. And in our commercial cultures it is all too easy, and actually is not helpful, to conclude that when profit is seen hovering at the door ethics inevitably flies straight out of the window. The second fact to be re-emphasized is that scientists and those who write about science are operating in an increasingly skeptical world, where a larger and larger number of people have populist aims, for which one feels a great deal of sympathy.[99] Whenever the press grants space to the claims of a whistle-blower, for example, there is a selection and a decision, for the press makes it possible for the whistle-blower to exist at all. That is undoubtedly a good thing, but it increases the imperatives on the scientists to be open and honest and on the press to be temperate and accurate.

But there are some other questions that we must ask at this stage, before we ask what might be done to improve the communication of science, its ideals and implications, to the public at large. First, we must ask what it is the public needs to know; second, what we can insist on from both parties to this task; and lastly, we return to the core question: What should be the relationship between the scientist and the journalist in this task, and should the journalist take on the role of critic?

The public's right to know about science and its implications is paramount. It needs to know, first, the hard facts of scientific discovery and their relationship to past and changing ideas. Second, it needs to know what are the current scientific and trans-scientific issues, the areas of concern and debate, especially as they relate to the impact of scientific ideas on those social and political issues on which the public will be voting or on which citizens should make their opinions felt. And third, the public needs to know about the actual nature of the scientific process, for this, as much as the content of science, should be comprehended. The patterns, the limits, the nature of discovery, the balance of certainty and uncertainty must be made explicit. What one hopes for here is that the methodology of science and the spirit of science should also be conveyed, along with their underlying relationship to the basic factual information. Of these three tasks, the media have traditionally done the first very well indeed. The second they are only just beginning to grapple with; and the third task practically no one has done at all, for journalists generally don't know how to, and scientists either have not bothered or, in some cases, have not wanted to be observed that closely.

Now given these essential tasks, what should we expect—indeed, what should we insist on—from both parties to this work? The answer is easy to state. We should insist that everyone be truly professional, with the word "professional" meaning three things already mentioned: a certain standard of competence, accepted and adhered to; an agreed ethic for proceeding; a responsibility as trustees *on behalf of the public and the society* which provides for all professional activity. This last is presently not only thoroughly unfashionable but often is not even considered to be part of the professional ethos. It was once, and in some professions—the medical and the legal professions in particular—it still is to some extent. But in this, one of the freest of free societies, there is a laissez-faire, noninterventionist feeling, so while professionals may deplore less than ethical behavior in others of their kind, they do very little about it. Yet doctors have been known to apply administrative sanctions against other doctors, and the public has certainly applied legal sanctions against doctors; scientists too have applied administrative sanctions against scientists who contravene the basic ethic by publishing false data. But certain exceptions apart, and their voluntary code notwithstanding, journalists and editors —and publishers—tend rather to shrug their shoulders with, "Oh, that is the market; that is the nature of the world."

Certainly there are watchdog organizations: The National News Council in the United States, the Press Council in Britain. A number of major metropolitan dailies have ombudsmen. The *Columbia Journalism Review* has been a highly regarded critical forum of the written word, although it has recently undergone a change of editorial policy which may reduce its critical role, to the detriment of the profession. All these are mechanisms for some degree of accountability through open public examination of contentious practices and procedures. While they do not possess weapons of sanction, they do possess weapons of embarrassment—if they choose to use them.

Nevertheless, despite these organizations and institutions, there does seem to be a double standard in operation here. It has been referred to, for example, in the pages of *Science* and also in *The Media and Business.*

In *Science* Dr. Stanley Garn specifically objected to what he saw as a dual system of ethics, one for the professionals of social science and another for journalists. The former are circumscribed by restrictions and ethics committees, in *their* gathering of behavioral data. By contrast, says Garn, a journalist, operating under the protective umbrella of the First Amendment and traditional press freedom, can behave in ways that have recently brought torrents of criticism on the heads of social scientists.[100]

In a similar vein, businessmen have railed against perceived hyprocrisy, as the introduction to *The Media and Business* makes clear:

> Businessmen are especially angered by what they think is a double standard utilized by the press. For example: If Boeing or Lockheed pays bribes to foreigners for contractual favors, that's viewed as a "no-no." If the press pays for a story, that's a "yes-yes."
>
> Inducing someone to commit a crime is a crime for most persons. But the press seems to think it is all right to induce a grand juror, for example, to break the law, or to induce someone to steal a document and provide it with a copy.
>
> The average journalist says that he or she will go to jail rather than reveal the source for a story if that source has asked for confidentiality. But the press cannot wait to publish the secrets of government and business.
>
> This perceived hypocrisy was described by one businessman . . . in the following manner: "What I got out of this [conference] and what I think disturbs me most, is the double standard between the press and business on matters of responsibility. I was, frankly, surprised to hear a television executive say that he wouldn't wait a day or two to go to press, or go on the air, with the story about an [CIA] agent, even though it might cost him his life, for fear of competition. Whereas I don't think the press would stand for it if a drug company rushed its product out quickly, for fear of competition.
>
> I heard it said that the press will protect confidentiality at almost all costs, whereas if a business tries to, it will be considered a cover-up.
>
> I heard it said that the press will one way or another buy information, while business will be castigated for questionable or sensitive payments.
>
> And finally, I heard, today, that the press reserves its right

to observe orders of the courts, even when none of us would tolerate the fact that the President of the United States raised that reservation. No businessman could say, 'I will reserve the right to observe an order of the Court.'

I think, in looking at business, perhaps the press could look at it from its own perspective of its engaging in competitive practices and perhaps the coverage might be a little more balanced."

It brought the house down with applause.[101]

What we have to ask ourselves is first, is there really a double standard? If so, is this perhaps the price we have to pay for a free press, and again if so, are we willing to pay it? That *some* price will have to be paid is quite clear, and there may have to be more than one. For Tom Stoppard the price paid for the small part that really matters is the volumes of trash encapsulated in the delectable series of junky headlines that Ruth, the main protagonist in his play, consistently flings out. But in the end too, she will admit, though she does so in tears, that yes, while the young journalist died for the women's page and the reporter's ego, and for the crossword and the racing results, somewhere "at the end of a long list" he died, she supposes, for the leading article, too. But is it worth it? Yes, the photographer Guthrie says, "Yes, it is. For information is light," and we have to agree. Information about anything, Stoppard says, is its own justification. So perhaps, in response to us, journalists will say that the only *demand* we can justifiably make is that there should be a free press. And that *that* is the only guarantee they can give us.

Nevertheless, in the context of this fundamental issue, I believe we also have to demand a new degree of responsibility. The media can fairly be asked to try to be accurate at all times. But in this new era, with so many important transscientific and social issues, the ethics of good journalism must apply with special force to the reporting of science and of scientific issues. The media can be asked to weigh again the right balance that must be struck in reporting, for example, the emergence of individual and collective activism on the one hand, and their "lazy" habit of turning for information only to those easy-access, "visible scientists" who may not be the most competent to comment on a particular scientific matter. The thalidomide affair should be a cautionary tale in this regard, and this brings me back to the question of criticism and advocacy I mentioned earlier.

There are several dimensions to this. It always was a difficult question as to how the readers of popular science could be helped to distinguish between the factual, certain portion of an article and the exciting, gee-whiz, uncertain elements. Now, unlike the good old days when there seemed little to disagree about, they must also be helped to judge between dissenting scientists. So the journalist has the added role of both presenting equitably and intelligibly the views of scientists who disagree with each other and helping the public understand and decide between them. In coming to terms with this situation journalists must decide (i) whether they will play this role, and (ii) what criteria they will apply to these matters. The journalist always did need to translate scientific information into terms that the reader (or viewer) could understand; now what he or she must also do is consciously offer help in the assessment of difficult trans-scientific issues—and to do *that* journalists are going to have to get very solidly educated in these issues.[102]

And what can be demanded of the scientist, who may seem to have come off rather lightly so far? In two words, candor and cooperation. Candor does sometimes hurt in the short run, but in the long run it generates real understanding. It must be an accepted task of the scientific community to educate and inform the journalists and the public. Given the present circumstances it would be smart for scientists to behave as they did at Asilomar: forthcoming, open, honest, and articulate. It would be wise for them, too, to be more charitable to those members of their own profession, and to those on the periphery who are honestly attempting to interpret science for society, in the hope that science will be truly assimilated into the culture.

Yet as public concern over recombinant DNA dies down, one begins to see a regrettable relaxation among many members of the scientific community. It is almost as if they had learned nothing from that episode; that, since the worst is apparently over, they feel free to get on with their jobs as if nothing had happened and certainly as if there were no lessons to be learned. True, there is the realization that an adequate science policy must be formulated, so relationships with those who make, administer, and determine this policy will have to be improved where they do exist and called into being where they do not exist. Nevertheless, there is, so far as I can gather, no corresponding conviction that the public has to be informed and brought along too. Once again the scientists seem about to commit the cardinal error of ignoring the fact that their profession sits squarely within the social matrix. It is this social matrix that helps determine the directions in which science will go and the social acceptance of its fruits, yet apparently scientists have still not fully realized that the contemporary social matrix demands that the desires of ordinary people must be respected and acknowledged.

If we were tempted to bypass the public and public communication, the nuclear accident at Three Mile Island in Pennsylvania brought the lesson home again. Over and over, through the news media the complaints of ordinary people came through—people who were clearly not willing to tolerate this kind of situation indefinitely, especially when they suspected not only that they had been misled by experts in the past but were, in the present, being fed soothing pap by the public affairs department of the utility firm. When initially they were also unable to get any clear information from federal sources, one could sense the anger rising. On the Sunday after the accident, in the 90-minute CBS news program *Sunday Morning,* various people who had to remain behind in one of the small towns near the reactor were interviewed. One of them, a fireman, said sadly and seriously, "If only someone would come down and explain to us, in layman's terms exactly what happened, exactly what is going on, exactly what we can expect, then," he said in effect, "we could cope and we could manage." The public's desire to be kept fully and honestly informed is going to intensify. There can be no return to those days when only a small band of experts had to influence a small band of influential politicians.

In one important way the scientist has to be more activist than the journalist and to change more. It is not the role of the media to help or to hinder the progress of science directly; it is the job of the media to report and explain it. And it is the job of the scientist to help the media in this task and to deal with curiosity, skepticism—but not, one hopes, cynicism or inquisitorial probing—on the part of the press.

Ideally, the relationship of scientist and journalist should be at arms length, but cooperatively at arms length. Nor should one demand of the other the impossible. Journalists should not ask that scientists be free with their data until they have at least gone through the preliminary assessment of the peer review. Scientists should not insist, or expect, either that journalists are there to help them with their public image or to disseminate their results. Most importantly, they should not identify the media with the public. By taking the media into their confidence, they open only one door to the public. The full job of educating the public requires, I think, further action from the scientific community *quite apart from its relationship with the media*, but this is a much larger task and beyond the scope of this essay.

New attitudes and new relationships are not easy to achieve. Science writer Victor Cohn once described the relationship of a scientist to the media in these terms: "Scientists are to journalists what rats are to scientists."[103] The journalist thinks that science is newsworthy and observes the scientist as the scientist does his animals—independently. But just as a rat's behavior can never be fully isolated from the environment in which the rat is living—as some scientists have found to their cost—so in assessing what is being done in science, a responsible journalist will weigh in the balance the attitudes and mores of the scientific profession, and their aims, against those of society. And it is very important that each not be taken in by the other's propaganda. For in the end, what we need to achieve is a situation of mutual understanding—an understanding (as Barbara Yunker has put it) of our "intense interrelationship in moving toward the common goal of good science done under ethical standards, watched by a crew of independent, hopefully intelligent observers who will tell the truth of what we see to the best of our abilities."[104] But that cannot be achieved unless, as Dr. Norton Zinder said, "scientists are willing to talk without anxiety and journalists are willing to listen without deadlines."

Yet beyond all this, I believe present circumstances require a new group of people, outside the scientific profession yet looking at it critically. The new role which has been created by the circumstances and social ambience in which science now operates dictates this, as does the very nature of scientific work which increasingly impinges on the autonomy of every individual. This new role has been recognized by such people as Dr. Lewis Thomas who, at the Rochester Conference, spoke of the need for a "highly professional, analytic, meditative scrutiny of how [scientific] work is done; why; whether good, bad, or trivial; and what is its inner meaning." It is not, Thomas pointed out, the same job as the daily reporting of science. It is, I believe, much more the role played by a Walter Lippmann in politics; and an F. R. Leavis (a famous British critic) in English literature; and a Donald Tovey in music. It is, in fact, "a critic" in the accepted, old-fashioned meaning of the term.[105]

Now we have long been used to having critics in music, literature, and the other arts. They have, when they served honorably, played a vital role in interpreting, mediating, and mirroring both the activity and the products. I used to believe that such critics would be redundant in science, an activity whose lifeblood is the criticism of peer review. This was too narrow a view, for it is clear to me now that no profession can truly be its own critic. It may have an internal methodology which is sustained by internal criticism, but few groups can look detachedly at themselves and their work. I see informed, independent assessors and interpreters of science playing as vital a role in the relationship of the enterprise with society and the assimilation of its ideas as their counterparts do in the arts and humanities.[106]

And I see another reason why such people will be necessary: the increasingly skeptical attitude of journalists. As one such person recently said to me, "I never approach any person I'm interviewing without asking myself, 'Why are they lying to me? What are they trying to hide?'" I queried this: "Scientists, too—talking about their discoveries?" He replied, "Absolutely. They rewrite history. I'll only believe them about a discovery or a fact if they take the time and trouble to take me right through the history of their subject—even if it takes two weeks of their time and mine."

Now there is a problem here. Journalists are in "the challenging business," of course. And it is also true that the degree of skepticism and questioning that journalists now bring to science is partly a reaction against an earlier time when that relationship was much too cozy, and journalists much too wondering. Thus, by some, to be credulous at all is now seen as both inappropriate and unprofessional. On the other hand, my friend's absolute attitude may be that of someone really ignorant about the scientific process. As the late Jacob Bronowski wrote:

> You cannot take the simplest statement in science without having to believe a lot of people. I know an awful lot of biologists and there are many subjects about which I would not believe a word they said. But when they start talking about how DNA is put together, then I know that they are telling the truth. We could not work without that tradition. ... That has been the quickening life force in science which has made it possible for people to have absolute trust in one another's statements.[107]

So if we are going to have journalists whose job it is to challenge everything, we also need to have writers about science who understand the tradition fully. But to accept—as I think we will have to—this suggestion for interpreters and critics of science is also to demand great responsibility from those who would undertake this critical task. For a critic's role has its own temptations: to do a hatchet job for the sake of doing a hatchet job; to become so narcissistic or intellectual that the end and the object becomes one's own ideas rather than those of the person or enterprise under review. These temptations are very real and have to be strongly resisted. Equally, it will be extremely difficult for the scientific profession to accept either the need for such critics or their existence. Yet to my mind there was a crucial humility in Lewis Thomas's suggestion which, if implemented, could be extremely valuable—but only insofar as the scientific profession came to trust the critics and opened the profession to them.

Finally, I want to make a few less radical suggestions. Scientists certainly do need to reach out to the public in ways they have not attempted before, but my first and most simple suggestion is to make better use of what is already to hand. For a start, many scientific institutions and universities have public information offices or public relations departments, and these have not always been used wisely or fairly. Public information people, so Gerald Delaney tells me, are secretly grateful for the boon of Watergate. For, he said, it illustrated all the things they had been warning against for years: covering up; stonewalling; not being candid; not adhering to the ethical standards. Individuals in public relations are, of course, loyal first of all to their institutions; that is why the press automatically distrusts them to a certain extent. But just as public information officers have a professional obligation to advise the press if, for example, they think that reporters are misguided on a story, or that an injustice will be done, so they also have a professional obligation to advise their institutions that autonomous operation is not feasible and that the consequences of this fact must be faced. Too often public information people are used either to prevent access, or to prevent the truth from being revealed. But their function must be to open the doors in *both* directions, and that is going to be painful.

This is merely one practical point about information. What about process? The workshop meetings that have already been started between scientists and science reporters should be extended not only in number but in personnel to include newspaper and television editors, sub-editors and directors. So far as the nature of the scientific enterprise is concerned, those are the people who probably most immediately need educating.[108]

Within the scientific community itself, the possibility of clearing-houses or places to which reporters might refer for orientation about continuing scientific activities or ambiguous scientific data have been discussed for some time. We could certainly have used one such on the question of drug testing and tranquilizers. This function could be undertaken by various scientific societies or institutions, and certainly this is one that should be implemented as soon as possible. There are models in existence. The Massachusetts State Legislature has devised a system for the dissemination of information which links legislators, experts and citizens themselves. One bonus from this system is the feedback from the public. Such an information pool devised by the various scientific institutions would provide a very valuable aid, not only to the media but to legislators too.[109]

One institution has tackled this problem in another way. The Center for International Environment Information has already produced one directory for journalists on energy, and in the fall of 1980 will be publishing a second, *Guide to Specialists on Toxic Substances*, produced with the help of a grant from the National Science Foundation. These directories list specialists who are the top U.S. experts in the fields and who have agreed to answer telephone queries from reporters. Some 2000 individuals are listed for toxic substances, drawn from government, industry, environmental organizations, labor unions and the academic community. The guides are offered free to newspapers, radio and television stations and provide an invaluable resource for journalists in these specific areas. The guides also supplement the service offered by the Scientists' Institute for Public Information, which maintains a roster of several thousand scientists, cross-referenced by area of expertise.

Next, within the institutions of science and the universities, some consideration should surely be given to a career structure in the interpretation and communication of science, for those people who deliberately might choose to take up the function of science interpreter/critic. They will not necessarily end up at the television station or the newspaper. Some indeed might wish to combine a career in research with one in writing, and there are institutions, for example, MIT, that have made a start in this direction. One reality here is that, except for a *very* few people, science writing is absolutely *not* a lucrative business. As things stand at present, the market dictates, and it is almost impossible to make a reasonable living only by writing serious *good* books or articles about science alone—basically because it takes so much time. Slowly, people other than science writers are beginning to realize this. Twenty years ago, I was told, the Sloan Foundation would never give money to people who were writing books about science if there was any expectation that these would be published by commercial publishers. Now the Foundation realizes that if this principle were totally adhered to, few good science books would ever be written, and in collaboration with Harper and Row it has established a program which has already produced some good literature about science. Much, much more is needed, and it would be nice to feel that eventually good communication of science will drive out bad communication simply through the sheer weight and wealth of material.

So far as I can see, the biggest single impact on the problem of mass science communication could be made through television, and this is where I hope to see a really substantial effort. Somehow, through private foundations, through public pressure on corporations and networks, or some other way, we must really begin to get good science programs in reasonable quantity available on the *commercial* networks, with ancillary book material available at the same time, and made under those conditions that will lead to proper analysis as well as brilliant television. There are rich veins of material and interest to be utilized, if only enough interest and money can be mobilized. Public television activities were mentioned earlier, and where the imagination and creativity of scientists can be combined with enough other resources, the results in this medium can be spectacular. It is a great pity that by far the greatest proportion of television specials sponsored by the big corporations are devoted to the arts. Certainly I wouldn't want to drop a single one of them, but what is the rationale that ignores thereby one enormous area of human creativity?

* * *

For one reason or another writing this essay has taken a very long time. During the many months of discussion, interviews, research and writing, both the material and the complexity of the issues threatened to overwhelm me. And if that were not enough, events too began to overtake me. For now we are seeing what can only be described as a remarkable "renaissance" of interest in science—an interest that has been quickly reflected in the printed word.[110] The items abound: *The New York Times* Science Section, a weekly feature appearing on Tuesdays; *Science 81*, a monthly subscription journal of science published by the American Association for the Advancement of Science; *Geo*, another monthly magazine of superb quality though expensive in price; *Omni*, a monthly with a strong editorial commitment to space—both inner and outer—that combines science and science fiction. Time Inc., has launched *Discover*, a monthly periodical aimed at that part of their existing audience already interested in the science, medicine and behavior sections of *Time* magazine, while the Hearst Corporation has a new format for *Science Digest*, a monthly magazine that has been around for nearly forty-two years.

Yet, while the new surge of interest is welcome, of itself it does nothing to diminish the seriousness of many of the questions which, as pervasive threads, weave in and out of any serious reflections on science and the media. I posed them at the beginning of this essay, and at the end they rise up to challenge me again. There is Murrow's: "Are we in an age in which intelligence may not be able to simplify truthfully?" "Can ordinary people be reached with complicated truths?" There is Tom Stoppard's question: "In a free society what is the price we must pay for truthful information? Is the only thing that a free press *can* guarantee is that it is free?" There are the queries of irritated scientists and business men: "Should not the press be held as accountable as it insists any other profession or institution should be?" These are the general questions within whose framework lie the more specific problems that relate to communicating science, and I cannot pretend to have answered them.

One thing can be said: the burdens must be shared, for the responsibility is a dual one. It does little good to conduct historical post mortems, and in any case, mere understanding is not enough. Maintaining a level of scientific literacy in the public is as difficult a task as doing science itself, and the media cannot do this alone.[111] Each party in this relationship, which is surely a symbiotic one, has to accept functions over and above those internal to each profession. Moreover, each must exercise these in the context of that extended ethic that lies at the heart of proper professional behavior. The new problems that are arising at the interface between science and society are demanding that at least some science writers must become more like responsible political commentators, adding analysis, judgment and criticism to their reporting. And these very same problems demand that scientists must lift their gaze from their narrow professional concerns and become aware of the dynamic matrix that surrounds and supports their work. Society has given scientists the most precious gift it could: the conditions—the possibility, even—to do science at all. Therefore, scientists have a strong reciprocal obligation, and I look forward to the time when, if problems do arise at the interface of science and society, no scientist will dare to say, "I don't want to get involved," for the sanction would be unfailing peer disapproval.

Notes and References

1. A. J. Meadows, "Conference on the Public Understanding of Science and Technology," 28–29 July 1976, sponsored by The Science Foundation (London) and organized by Dr. Maurice Goldsmith, unpublished proceedings.
2. Personal interview, 7 April 1978.
3. This justification has never been more beautifully encapsulated than in Warren Weaver's "Report of the Special Committee on X," *Science*, v. 130, 20 November 1959, pp. 1390–1391.
4. Personal interview, 7 April 1978.
5. Ibid.
6. Personal interview, 22 October 1979.
7. H. Peter Metzger, "Advice from the Outside: Why the Press (Still) Doesn't Understand Us," presented by INFO 75, Topical Conference on Nuclear Power and the Public, Atomic Industrial Forum, Inc., Los Angeles, California, 5 February 1975.
8. See, for example, Barry Bloom, "Health Research and Developing Nations," *Hastings Center Report 6*, December 1976, pp. 9-12.
9. Alvin Weinberg defined these explicitly as "issues at the interface between science and politics," or "between science or technology and society." See, for example, his contribution to the discussion on the responsibility of scientists, "Science and Trans-Science," in the CIBA Foundation symposium, *Civilization and Science*, Amsterdam: Elsevier, 1972, pp. 71, 105.
10. I myself took part in the Astor Conference, held at the University of Sussex, September 1973, and I have made extensive use of my own notes taken at the time. I must also acknowledge the help provided by the ideas of my colleagues at that meeting.
11. National Academy of Sciences' Academy Forum on Recombinant DNA, March 1977. The book based on the Forum is *Research with Recombinant DNA*, Washington, D.C.: National Academy of Sciences, 1977.
12. David M. Rorvik, *In His Image: The Cloning of a Man*, Philadelphia: Lippincott, 1978.

13. Edward R. Murrow, "Television and Politics," British Association Granada Lectures on Communication in the Modern World, in *Dons or Crooners?* Granada TV, 1959, pp. 64–66.

14. Metzger, for example, is relevant again here.

15. Rod MacLeish, CBS-TV Evening News, 4 March 1978.

16. John Ziman, in a radio broadcast, "Seeing Through Our Seers," reprinted in *The Listener*, 24 June 1976, p. 794.

17. The first article by the Insight Team appeared in *The Sunday Times* (London), 17 February 1963.

18. *The Sunday Times Magazine*, 26 October 1969.

19. "Long-Term Forecast," *Newsweek*, 23 January 1978, p. 74.

20. *Time*, 25 July 1969, p. 10.

21. *Time*, 13 January 1975, p. 60.

22. For example, consider this story, run by a Bronx, New York, paper which was, I am sure, consciously committed to the importance of informing its readers about science. On Wednesday, 4 October 1978, *City News*, in its feature, "Science of Yesteryear," under the headline, "The Theory of Evolution and a Man Named Darwin," printed these two opening paragraphs:

> One hundred and forty years ago, a 29-year-old naturalist sat down in his London apartment to read an essay "for amusement." His eyes moved over the words and an idea formed in his mind. The reader's name was Charles Darwin, and his idea was the theory of natural selection—the mechanism of evolution.
>
> No other theory has shaken scientific thought more than Darwin's: that the environment could modify living organisms, and the modifications could be passed on to the next generation.

Darwin is being credited with Lamarck's theories—which fundamentally were 180 degrees opposite his own. One wonders whether reporter or editor had ever checked out Darwinian theory with any competent biologist or science historian.

Errors can sometimes be serious, or sometimes merely embarrassing to a careful science reporter. An example of the latter kind of error slipped into a recent *Time* magazine story on the satellite rendezvous with Jupiter (12 March 1979, p. 86). The picture of the Great Red Spot sent back by the satellite was printed upside down, not because the science editor didn't know that Jupiter's spot is in the planet's southern hemisphere, but because the picture editor thought that the picture "looked better" the other way up!

23. The sensational, eye-catching headline is certainly a persistent problem from the point of view of the scientific and academic community, and it occurs even in such respected magazines as *Saturday Review* and *Harper's*. An exchange which was recently recorded in the *Chronicle of Higher Education* ("Dean, Under Fire for Article, Says Magazine Sensationalized It," 6 March 1978) illustrates the differing perspectives of scholar and editor: Theodore

L. Gross, at that time the Dean of Humanities, City College, New York, had written a lengthy essay on the policy of open admissions at colleges and its effects. Thinking that academics should write for audiences outside their specialties, he was pleased when *Saturday Review* accepted his article. But when it ran in the 4 February 1978 issue, he took great exception to the new title, "How to Kill a College: The Private Papers of a Campus Dean," and to the lurid graphics showing the university building being stabbed by a hunting knife, with blood coming from one of the upper-story windows. His own title had been "Open Admissions: A Confessional Meditation." He protested to the editor, both because he was not consulted about the title or illustrations and because of the actual wording of the title. But, the editor replied, magazines and newspapers rarely consult with authors on these matters, so there was nothing new about that. And, his letter went on,

> Our job as editors is to present articles in the most arresting fashion possible, without alienating our readers. Recent increases in our circulation indicate that our readers endorse our decisions. I am willing to bet that our title and graphics will attract more readers to your article than a tamer title and graphics would do. Needless to say, I thought your article was first rate. I think your misgivings about our handling of the piece were unjustified.

This example illustrates not only the problem but the nature of the different allegiances and the additional difficulty of mutual misunderstanding of the other's aims—or of deliberately choosing to ignore them. Dr. Gross felt that those graphics and that headline, purely and solely dictated by the desire to "grab" readers and increase circulation, were unacceptable. Most academics will sympathize and would have—indeed, do have—similar misgivings about popular communication: their views may be misquoted in a headline; or distorted in a story; or partially used in various contexts by other authors, or anyone else, so making them appear to say what they do not believe or have not said.

Headlines were Dr. Adele Simmons' concern also. Her article in a recent *Harper's* (March 1979) dealt with Harvard's curriculum reform, and when she saw it in galley proof the title had been changed to "Harvard Flunks a Test," a title which, as she wrote to the editor of *Harper's*, "misrepresents the article, causes me profound embarrassment, and wrongly prejudices the reader" (*Harper's*, April 1979, p. 91). As president of Hampshire College, she would indeed be embarrassed by seeming to characterize Harvard's reform as a failure. But there was very little that she could do except protest, and these situations are unlikely to change. For the pressures of the market place will continue to operate, and so too will the misgivings of the academics who communicate with a wide public.

24. Tom Stoppard, *Night and Day*, London: Faber & Faber, 1978, p. 91.

25. Daniel Greenberg, *New England Journal of Medicine*, 24 November 1977, p. 1188.

26. Roger G. Shepherd and Erich Goode, "Scientists in the Popular Press," *New Scientist*, 24 November 1977, p. 482.

27. Rae Goodell, *The Visible Scientists*, Boston: Little, Brown, 1977.

28. As recounted by Robert Reid in "Television Producer and Scientist," *Nature*, v. 223, 2 August 1969, p. 455.

29. For a full discussion of this, see Ron Powers, *The Newscasters*, New York: St. Martin's Press, 1977; see also John J. O'Connor, "Independent vs. Editorial Control," *The New York Times*, 25 November 1979.

30. Reid, p. 455.

31. For example, Jay McMullen, a producer with CBS news, told me that he has been able to do only two documentary programs on science, plus two science news segments, in over 25 years with the network.

32. Carl Sagan, "In Praise of Science and Technology," *New Republic*, 22 January 1977, p. 21.

33. Marcel Chotkowski La Follette, "Wizards, Villains, and Other Scientists: Science Content of Television for Children," unpublished study. (Copies can be obtained by writing to the author at Room 20B-125, MIT, Cambridge, Massachusetts 02139.) This study attempted to assess both the quality and quantity of explanation about science, and the depiction of science, on programs which were specifically designed for children or so placed in the schedule that children in significant numbers would see them.

34. For example, see the letter by Richard L. Garwin, "*60 Minutes* on Particle Beam Weapons," *Bulletin of the Atomic Scientists*, February 1979, p. 50.

35. As originally planned, the first *Universe* program was to cover the history of the universe, with two billion years compressed into a commercial television time slot of one minute and 15 seconds; a report on the likelihood of earthquakes in Los Angeles; an examination of recent satellite photographs of the far side of the moon; and a report on the search for clues to the cause of multiple sclerosis. These were to be aired in July and August 1980.

36. W. R. Klemm, *Discovery Processes in Modern Biology*, Huntington, New York: Robert E. Krieger, 1977, p. x.

37. "Medicine and the Media: Ethical Problems in Biomedical Communication," a symposium presented at the University of Rochester Medical Center, 9–10 October 1975, unpublished proceedings.

38. *The Washington Post*, 24 May 1965, p. 1.

39. Rochester Conference, unpublished proceedings.

40. Gail McBride, "The Sloan-Kettering Affair: Could It Have Happened Anywhere?" *Journal of the American Medical Association*, v. 229, 9 September 1974, p. 1391.

41. Personal communication.

42. Quoted in McBride, pp. 1409–1410.

43. For an extended discussion of this entire episode, see June Goodfield, *The Siege of Cancer*, New York: Random House, 1975. See also Joseph Hixson, *The Patchwork Mouse*, Garden City, New York: Anchor Press, 1976.

By the time I came to write *The Siege of Cancer*, Gail McBride's article had been published, and I was able to make extensive use of her excellent work. I needed to, because I had been in the unusual position of having been very close to the event. Summerlin had applied the felt tip pen to the mice at 7 o'clock on that Tuesday morning (26 March 1974) and was temporarily suspended, pending investigation, at noon of the same day. But between the two events he gave me a long interview about his work (though not, of course, about his deception!). The other popular account of this affair, well researched, thoroughly detailed, and designed quite specifically for the mass audience, was Joseph Hixson's *The Patchwork Mouse*. In the course of some extremely painstaking work, Hixson rightly took issue with his fellow science writers and journalists about this very same point, namely, the way they rushed to their conclusion without proper examination of the evidence. Unfortunately, he failed to point out that there was one truly outstanding exception—the Gail McBride article—although it is clear that he was aware of her article. In fact, at the very beginning of his book he lists seven articles, permission to use quotations from which he gratefully acknowledges. Gail McBride's article is actually third down the list but, surprisingly, it is the only one with the author's name omitted. So Sir Peter Medawar, reviewing Hixson's book for the *New York Review of Books* (15 April 1976) missed the McBride article, too, and so did the public.

44. Goodfield, pp. 228–229. It is interesting to compare this statement with conclusions about the Summerlin affair. For example, see Barbara J. Culliton, "The Sloan-Kettering Affair: A Story Without a Hero," *Science*, v. 184, 10 May 1974, p. 644; and Barbara J. Culliton, "The Sloan-Kettering Affair (II): An Uneasy Resolution," *Science*, v. 184, 14 June 1974, p. 1154. The real point here is the extent to which people's personalities were presumed to have affected the tragic episode, as in the Summerlin affair, but which were presumed to be irrelevant in the Rosenfeld affair. I think the inconsistency should be highlighted.

45. William Bennett and Joel Gurin, "Science that Frightens the Scientist," *Atlantic Monthly*, February 1977, p. 43.

46. Michael Rogers, "The Pandora's Box Congress," *Rolling Stone*, 19 June 1975.

47. Stuart Auerbach, "And Man Created Risks," *The Washington Post*, 9 March 1975.

48. Richard Hutton, *Bio-Revolution: DNA and the Ethics of Man-Made Life*, New York: New American Library, 1978.

Note added in proof: Since this monograph went to press, Rae Goodell has written an outstanding article which analyzes both the press involvement in recombinant DNA and demonstrates how fashion has once more influenced the media's attitude. She shows how, over the years since 1974, initial critical questioning of the political and safety issues has been displaced by enthusiastic journalistic pieces on the industrial aspects of the question. Thus the press has once more uncritically followed the "agenda of the scientific community" rather than developing its own. See Rae Goodell, "The Gene Craze," *Columbia Journalism Review*, November/December 1980, pp. 41–45.

49. Paul Berg et al., "Potential Biohazards of Recombinant DNA Molecules," *Science*, v. 185, 26 July 1974, p. 303.

50. Hutton, pp. 52–53.

51. Ibid., p. 83.

52. Rae Goodell and June Goodfield, "Rorvik's Baby," *The Sciences*, v. 18, September 1978, p. 6.

53. André E. Hellegers, "Book on Cloning Is a Hoax," *The Washington Post*, 4 June 1978, p. B3.

54. Barbara J. Culliton, "The Clone Ranger," *Columbia Journalism Review*, July/August 1978, pp. 58–62.

55. For scientific evidence on this topic, see Robert Gilmore McKinnell, *Cloning*, University of Minnesota Press, 1978, and, most importantly, a review of this book by Clement L. Markert, "Nuclear Transplantation," *Science*, v. 202, 22 December 1978, p. 1275–1276.

56. Bernard Dixon, "Rorvik—End of the Line," *New Scientist*, v. 80, December 1978, p. 914.

57. *Publisher's Weekly*, 13 February 1978, p. 113.

58. Sharon Churcher, "Baby Born Without a Mother," *The New York Post*, 3 March 1978, p. 1.

59. Ibid.

60. Leon Jaroff, quoted in "All About Clones," *Newsweek*, 20 March 1978, p. 68.

61. Robert Cooke, "They're Just Cloning," *The Boston Globe*, 8 March 1978, p. 13.

62. Harriet van Horne, "Clone Story Inspires Doubt and Horror, Not Wonder," *The New York Post*, 10 March 1978.

63. "Rorvik is a fraud and a jackass," quoted in "All About Clones," *Newsweek*, 20 March 1978, p. 68.

64. *Tomorrow Show*, 5 April 1978.

65. *Today Show*, 6 March 1978.

66. Culliton, p. 60.

67. Bruce Page, "London Diary," *The New Statesman*, 14 April 1978. Dr. Taylor has since written a book explaining how Geller deceived him.

68. Phillip Knightley, "Why Mr. Rorvik Is a Fraud," *The Sunday Times*, 23 April 1978.

69. Judith Randal, "The Cloning Controversy," *The Progressive*, May 1978, p. 11.

70. Rowena Farre, *Seal Morning*, New York: Scholastic Book Services, 1972.

71. See the introductory material by Leonard C. Lewin in *Report from Iron Mountain on the Possibility and Desirability of Peace*, New York: Dial Press, 1967.

72. Clifford Irving, *The Autobiography of Howard Hughes*, McGraw-Hill, publication cancelled.

73. Personal interview, 21 April 1978.

74. Judith Randal, "The First Cloned Baby? Despite Denials, It Is Possible," *The Washington Post*, 12 March 1978, p. C3.

75. Sissela Bok, "Lying," *The New York Times*, 20 April 1978, p. 23.

76. Ibid.
   In her article, Bok also comments:
   Lying by public officials is now so widely suspected that voters are at a loss to know when they can and cannot believe what a government spokesman reports, or what a candidate says in campaigning. . . . Two years ago, 69 percent of the respondents to a national poll agreed that this country's leaders have consistently lied to the people over the last 10 years. And over 40 percent agreed that most politicians are so similar that it doesn't really matter who is elected.
   Many refuse to vote under such circumstances. . . . Once trust has eroded to this extent, it is hard to regain. Even the most honest public officials then meet with suspicion. And in times of national stress when problems require joint efforts—problems, for example, of preparing for energy shortages or for inflation—the cynicism and apathy that greet government calls to common sacrifice are crippling.

   When this feeling extends to other important institutions and professions, and indeed to the public as a whole, then I believe that the country is well on the way to losing its collective self-confidence.

77. Lippincott's rationale was also put under a microscope in an illuminating article by Nat Hentoff, "The Case of the Pecksniffian Panderer-Publisher," *The Village Voice*, 2 October 1978, p. 57.

78. "Science News, Science Fiction and Enticing Readers into DNA," *Hastings Center Report*, v. 8, December 1978, p. 2.

79. Howard Means, "George Asmann Died Alone: And Now the First Victim of DNA Research at Fort Detrick Is Becoming a Non-Person," *Washingtonian*, September 1978.

80. *Hastings Center Report*, p. 20.

81. See Barbara J. Culliton, "Cloning Caper Makes It to the Halls of Congress," *Science*, v. 200, 16 June 1978, p. 1250.

82. The Insight Team of *The Sunday Times* (London), *Suffer the Children: The Story of Thalidomide*, New York: Viking Press, 1979.

83. Ibid., pp. 139–140.

84. This concept is discussed by Rae Goodell in *The Visible Scientists*.

85. Personal interview, 10 September 1979.

86. *Suffer the Children*, Chapters 2,3,4,7, and 9.

87. Ibid., p. 41.

88. Ibid., p. 123.

89. This was made explicit at a seminar held in 1977 and detailed in Howard Simons and Joseph A. Califano, Jr., *The Media and Business*, New York: Random House, 1979, p. 38. Discussing a case study which involved a hypothetical problem analogous to thalidomide and whether to publish warnings, one newspaper editor said:

> I just wanted to say on the area of the drug question—while I agreed entirely with all my colleagues about this story, and the way it ought to be checked—I *am* slightly uneasy about the idea that there is something terrible—and everybody is very defensive here—about publishing this story about the side effects of drugs. Perhaps this will provoke a hostile reaction, but I think we're overlooking the very positive aspects of newspaper publication, of which the most important is feedback from people.
>
> Just take the instance of thalidomide. In the thalidomide case, there were side effects of peripheral neuritis when this drug was taken. They were not reported. Newspapers were frightened by it. They wanted to be cautious. Doctors were persuaded to write favorable papers. There was a whole series of events like that. So the warning signals of peripheral neuritis, which was the first indication that we were about to produce the biggest drug disaster of all time, never got into the newspapers. Therefore, other people with peripheral neuritis never got in touch with their doctors, and the whole thing multiplied.
>
> In June 1960, an Australian doctor discovered that thalidomide caused birth defects. He sent an article to a medical journal. It wasn't printed. Pregnant women kept on taking this drug, and a vast number of deformed babies were born, conceptions which might have been prevented if that doctor had felt he could trust the newspaper.
>
> The drug was not withdrawn until [the end of] 1960. International publication of that story in a sensible way would have told people in Germany that things they didn't know were going wrong

were going wrong and somebody would have put two and two together. I think the press is being too defensive this morning.

90. Now at Charing Cross Hospital

91. Now in charge of the Drug Information Unit at Nine Wells Hospital, Dundee, Scotland.

92. The scientific bibliography at the end of *Suffer the Children* is most comprehensive with regard to thalidomide, but some early studies on other drugs are also cited.

93. *Suffer the Children*, pp. 179-80.

94. Ibid., p. 181.

95. Ibid., p. 203.

96. Ibid., pp. 224-225.

97. Ibid., p. 238.

98. Murray Rosen, "*The Sunday Times* Thalidomide Case: Contempt of Court and the Freedom of the Press," London: Writers and Scholars Educational Trust, 1979.

99. Many articles have been written recently about this phenomenon. For one that specifically discusses this in relation to science, see Frank Trippett, "A New Distrust of the Experts," *Time*, 14 May 1979, pp. 50-51.

100. Stanley M. Garn, "Social Science Research Ethics," Letter to the Editor, *Science*, v. 206, 30 November 1979, p. 1022.

101. Simons and Califano, p. xv. Many of the issues raised in this essay come up in *The Media and Business* in a variety of forms. The first case study (pp. 3-65), which discusses a hypothetical situation about a commonly used drug, with nasty side effects possibly surfacing, is especially valuable in the light of the thalidomide issue. Indeed, as I read this book, I felt that one could substitute the word "science" for "business" throughout, and most arguments and comments would still apply. Other material relevant to these points can be found in Herbert J. Gans, *Deciding What's News*, New York: Pantheon Books, 1979, and in Tom Margold, "The Witness and the Cheque Book," *The Listener*, 28 June 1979.

102. The pitfalls are many. For example, see the illuminating and instructive exchange in *The Listener* (22 March, 19 April, 31 May, 5 July, 12 July 1979) in connection with John Maddox's program on BBC Radio 4, *Scientifically Speaking*. His technique and format drew criticism because, it was said, Maddox was acting as a "publicist" for science, displaying a "simple faith in the revealed majesty of science" rather than giving a balanced account of socioscientific situations.

103. Rochester Conference, unpublished proceedings.

104. Ibid.

105. Critics may be even more necessary in the future than now. A

recent article in the *New England Journal of Medicine* ("Gene Cloning by Press Conference," 27 March 1980, p. 743) noted that some scientific data are being released at press conferences called by scientists who are by-passing the usual critical analysis by the community of science. Why do they do it? Extreme competitiveness is one reason, especially in a field, like recombinant DNA, which has enormous commercial potential. But how can journalists correctly assess the worth of work which has not been subject to peer review?

106. See also Tim Robinson, "Interpreters of Science," *New Scientist*, 4 October 1979, p. 53.

107. Jacob Bronowski, *The Origins of Knowledge and Imagination*, New Haven: Yale University Press, 1978, p. 131.

108. I mentioned earlier the American Cancer Society Seminars for Writers as one such workshop that might benefit from having the "gatekeepers" attend as well as the reporters. It should, however, be noted that many science writers are backing away from this meeting. The reasons are complex. One suggestion from the press is that in the absence of real success in cancer research, the ACS is doing more of a "lobbying" job for cancer research and less of an educational one. See E. Edelson, "The American Cancer Society for Science Writers—Who Gains More: The Writers or the Society?" *National Association of Science Writers Newsletter*, v. 26, September 1978, p. 1.

109. There can also be problems in this information transfer. In a recent article, Dr. Barry Bloom has shown how the National Cancer Institute has interpreted the Freedom of Information Act "as a mandate to submit data directly from the computer to the regulatory agencies, to publish the material in the *Federal Register*, and to release it simultaneously to the public as well." This is, he shows, a system of transference of data from animal tests to computers and then press releases without the mediation of a human brain—a dangerous omission, for science "does have an obligation to reveal the limits of knowledge" as well as provide information. And, as he insists, "data are not information." See Barry R. Bloom, "News About Carcinogens: What's Fit to Print?" *The Hastings Center Report*, 4 August 1979, pp. 5–7.

110. See, for example, Arlie Schardt, with Nancy Stadtman and Mary Lord, "The Science Boom," *Newsweek*, 17 September 1979, p. 104; and Dava Sobel, "The New Frontiers of Science Are in Print," *The New York Times*, 7 October 1979.

111. Some scientists are aware of this. See, for example, Edward Wenk, Jr., "Science, Engineers and Citizens," *Science*, v. 206, 16 November 1979, p. 771.

About the Author

June Goodfield, now a Senior Research Associate at the Rockefeller University, was trained initially as a zoologist. She received her doctorate in the History and Philosophy of Science at the University of Leeds, she has taught at a number of English and American universities, and she is a frequent lecturer on science and its ethical and human implications. In 1977 she was the Phi Beta Kappa speaker at the 143rd National Annual Meeting of the American Association for the Advancement of Science, and she is currently Adjunct Professor, Cornell University Medical College. She is the author of *The Growth of Scientific Physiology, Courier to Peking, The Siege of Cancer, Playing God: Genetic Engineering and the Manipulation of Life,* and most recently, *An Imagined World: A Story of Scientific Discovery.* She is the co-author of *The Fabric of the Heavens, The Architecture of Matter,* and *The Discovery of Time,* and she has also written and directed a number of scientific films. Deeply committed to the public understanding of science, she has spent her entire career working at the interface between science and the humanities.